AF784198

CYBER-PHYSICAL POWER SYSTEMS STATE ESTIMATION

CYBER-PHYSICAL POWER SYSTEMS STATE ESTIMATION

ARTURO S. BRETAS
University of Florida, USA

NEWTON G. BRETAS
University of Sao Paulo, Brazil

JOAO B. A. LONDON Jr.
University of Sao Paulo, Brazil

BRENO E. B. CARVALHO
UNISATC University Center, Brazil

ELSEVIER

Elsevier
Radarweg 29, PO Box 211, 1000 AE Amsterdam, Netherlands
The Boulevard, Langford Lane, Kidlington, Oxford OX5 1GB, United Kingdom
50 Hampshire Street, 5th Floor, Cambridge, MA 02139, United States

Library of Congress Cataloging-in-Publication Data
A catalog record for this book is available from the Library of Congress

British Library Cataloguing-in-Publication Data
A catalogue record for this book is available from the British Library

ISBN: 978-0-323-90033-1

For information on all Elsevier publications
visit our website at https://www.elsevier.com/books-and-journals

Publisher: Joe Hayton
Acquisitions Editor: Graham Nisbet
Editorial Project Manager: Madeline Jones
Production Project Manager: Poulouse Joseph
Cover Designer: Greg Harris

Typeset by SPi Global, India

Contents

CHAPTER 1

State estimation in electric power systems

This chapter aims to provide an overview of this book, highlighting the structure and motivation that led the authors to accomplish it by organizing the following sections:

- *Section 1.1 Introduction*: Briefly describes the value of the state estimation process for real-time operation of Electric Power Systems (EPSs), presenting concepts and the general nature of this process, contextualizing its development and the need for evolution to keep in line with the new trends and current technologies.
- *Section 1.2 Book Organization*: Describes its structure.

1.1 Introduction

1.1.1 Historical context

At the end of the 1960s, significant changes were made in the philosophy of the operation of EPSs. These were due to the blackouts in the East Coast of the United States, which showed the relevance of issues related to the security of the operation of EPSs. At the same time, the combination of two factors, the difficulty to operate EPSs (increasingly interconnected) and technological advances in computing and telecommunication, allowed the development of functions related to security monitoring and analysis, giving rise to the so-called Security Control or simply Real-Time Operation (Wu et al., 2005).

The initial step for real-time operation is to estimate the current state of the EPS and then determine the appropriate control actions. Once the topology and the parameters of the electrical network have been known, the operational state of an EPS, operating in a sinusoidal steady state, becomes a function of the complex voltages in its buses, since from them it is possible to determine all the other electric quantities of interest (currents and power flows). Thus in this context, the complex voltages in the EPS buses are simply called EPS state variables.

Cyber-Physical Power Systems State Estimation
https://doi.org/10.1016/B978-0-323-90033-1.00002-0

Due to the large size of the EPSs, their state variables are determined by means of telemetry systems, i.e., remote measurements, which are subject to a series of noises due to, for example, the different measurement accuracy of the transformers of the measuring instruments, the secondary circuit of these transformers (wiring and load), and the various measuring devices (transducers, converters, etc.) (Abur and Expósito, 2004). Thus in order to obtain more reliable values for the state variables as required for real-time operation, the measurements must be filtered. In order to perform this task, the control centers have a set of computer programs that are responsible for executing the so-called power system state estimation (PSSE) process.

From the works published by Schweppe et al. in the late 1960s and early 1970s (Schweppe and Wildes, 1970; Schweppe and Rom, 1970; Schweppe, 1970; Schweppe and Handschin, 1974), which outlined various concepts and the general nature of the problem, PSSE became the subject of numerous research. In works such as those published in Coutto Filho et al. (1990); Monticelli (1999); Abur and Expósito (2004), a vast bibliographical revision starting from 1968 is presented.

Currently, the PSSE process has already been consolidated as a basic procedure necessary for real-time operation of EPSs. However, due to the development of new equipment and electrical devices associated with the so-called smart grids (advanced measurement structures, intelligent meters, synchronized phasor measurements, etc.), the paradigms of the operation of EPSs expand and require the PSSE process to evolve to keep in line with the new trends.

In view of the previous, in order to work with the operation of EPSs, understanding the problems and suggesting solutions, the professionals of the area must have the theoretical background necessary to understand the functioning of the entire PSSE process. In this sense, it is important to highlight that, although several scientific articles have dealt specifically with several problems related to the PSSE process, relatively few published books cover all the stages involved in this process.

This was the main motivation for writing this book, which aims to present, in a didactic way and with the depth necessary for the training and specialization of electrical engineers, all stages of the PSSE process, based on several scientific articles published during the last decades dealing with specific issues in this process.

1.1.2 The process of PSSE

The first stage of the conventional PSSE process consists in obtaining the system topology with the correct location of the meters available in the so-called bus-branch model (which corresponds to the single-line diagram of the system), which is performed from logical measurements, obtained continuously consisting of status of switches and circuit breakers, as well as information about the type and location of the meters installed in the system (information modeled at the bus section level, that is, in the physical representation of the elements of the system).

Once the system topology is obtained in the bus-branch model, the second step of the process is to verify if it is possible, through the available set of analog measurements[a], to determine all system state variables. If so, the system is said to be observable. Otherwise, the lack of measurements can be met by pseudo-measurements[b], through which the entire system will become observable.

Since the system is observable, from the topology of the system and its parameters[c] stored in the database of the control centers, as well as the set of analog measurements available, the state estimation itself will be carried out in Step 3 of the process.

It is worth mentioning that the state estimator used in Step 3 can be dynamic or static. In the case of the dynamic estimator[d], the variations of the quantities of interest, relative to the variable "time", are considered in the network modeling. On the other hand, in the static estimator, the network model used is static, translating into a "photograph" of the system at a fixed time-instant. Thus the mathematical behavior of the electric system is translated using only nonlinear algebraic equations, without the use of differential equations. This book will focus mainly on the static state estimator, since it is the most used in current real-life applications. However, Chapter 8 will be dedicated to dynamic state estimation.

Because the analog measurements used in the PSSE process are not exact, the estimation of the unknown state variables will be also inexact. It can be said then that the PSSE process consists of finding the best estimate of the

[a] Those performed continuously and usually consist of active and reactive power flows in branches, active and reactive power injections, and voltage magnitudes in the buses.

[b] Pseudo-measurements are data of load forecasting, generation forecast, historical data, etc. that are part of the database of the operation centers.

[c] Impedance of transmission lines, position of transformer taps, etc.

[d] A broad literature review on dynamic estimators can be found in Shivakumar and Jain (2008), Coutto Filho and Souza (2009), and Zhao et al. (2019).

unknown state variables. For this, of the many existing statistical criteria, the one that has been used the most in the PSSE process is the Weighted Least Squares (WLS), which was originally formulated in Schweppe and Wildes (1970).

When the errors in the measurements are Gaussian, the classical WLS Estimator works very well, failing, however, in the occurrence of one or more gross errors (GEs[e]). Among the causes of GEs, we can highlight "analog-to-digital" conversion errors and errors in the telemetering communication channels, or even cyber-attacks (Liu et al., 2009).

In the case of single GE, that is, when only one measurement has GE, or multiple noninteracting GEs[f], the classical WLS estimator associated with techniques for processing GEs from the residual analysis perform well. However, it may fail in the following situations (Abur and Expósito, 2004): GEs associated to measurements with low redundancy[g]; multiple interactive GEs; and GEs in measurements that are highly influential, that is, they attract the convergence of the state estimation process, traditionally called leverage point measurements (Monticelli, 2000). One should observe that although some works on such refer to error, they want to say residual, because the classical WLS works with residual only and not with error, which are different quantities (Bretas et al., 2013).

It should be noted that due to the previously cited problems, statistically more robust state estimators were applied in EPS, such as the Weighted Least Absolute Value (WLAV) method (Irving et al., 1978; Kotiuga and Vidyasagar, 1982). The WLAV estimator was shown to be more robust than the WLS associated with residual analysis techniques in the presence of single and multiple GEs. However, the WLAV also fails in the occurrence of GE in one or more measures of leverage point (Falcão and Assis, 1988). In order to overcome the problems caused by the measurements classified as leverage points with GEs, the estimator based on the Weighted Least Median of Squares (WLMS) method was proposed for EPS (Mili et al., 1991). This estimator was the first proposed statistically robust estimator that can identify GEs in measurements classified as points of

[e] In general, we say that a measure carries GE, when it deviates from its "true" value of at least three times its standard deviation (Van Cutsem et al., 1984).

[f] Definitions and concepts pertaining to the types of GEs in measurements, as well as to the residual analysis of the measurements, will be detailed in the course of this book, more precisely in Chapter 6.

[g] Redundancy, a topic that will be dealt with in detail throughout this book, refers to a surplus of measurements in relation to the number of state variables to be estimated.

leverage. However, such an estimator requires a combinatorial search, making it impracticable for real-time application in large EPS (Falcão and Arias, 1994; Monticelli, 2000).

Due to the practical unfeasibility of the WLMS estimator, and due to the simplicity of the formulation and the ease of computer implementation of the WLS estimator, associated to the analysis of the residuals of the measurements, the latter is the estimator that is most used in real-life applications and most studied in academia.

After the considerations in the previous paragraphs, in the fourth stage of the PSSE process, GEs are processed, that is, the procedure for detecting, identifying, and eliminating GE-carrying measurements (or the effect of these measurements). If any measurement is detected with GE, after processing of Step 4, Step 3 is performed again to obtain state variables without the effect of the erroneous measurements. The PSSE process ends when, after an estimate, no measurement is detected with GE. One should observe again that great majority of previously presented and implemented real-life solutions are using the residuals to detect GEs.

The classical formulation of the PSSE process is based on the assumptions that the network configuration (obtained in Step 1) and the parameters used in the models representing the elements of the electrical network are correct. However, these assumptions are often not true. Thus in addition to the GEs, the PSSE process is still subject to two other types of errors. These are error caused by erroneous information of some parameter of the electrical network, known as parameter error (PE), and error due to erroneous information about the status (open or closed) of a switch and/or circuit breaker in Step 1, known as topological error (TE). The effect of TEs and PEs on the PSSE process is drastic, usually intolerable, and several researches have been developed to address these errors.

Although the generation, transmission, and distribution of electric energy usually occur through three-phase systems, in the classical formulation of the PSSE process it is considered the hypothesis of a balanced system. That is, it is considered that the loads are balanced, and the networks are symmetrical, where the electric quantities in the phases have the same magnitude, but with a phase angle difference of 120° between them. Thus most of the studies developed in the context of PSSE make use of the model per-phase, or single-phase (positive sequence model), suitable for most transmission systems, which is the focus of this book.

It should also be pointed out that in the classical formulation of the PSSE process only the unsynchronized logical and analog measurements provided by the Supervisory Control and Data Acquisition (SCADA) system are considered.

1.2 Organization of the book

In order to ease the understanding of all stages of the PSSE process, as well as some problems related to this process and the proposed solutions already presented, this book is divided into nine chapters, where the Introduction corresponds to the first one. In order to make the text more clear, page footers with explanatory notes will be used in all chapters, defining or explaining the meaning of terminologies common to the area to which they refer to.

The chapters were elaborated following a logical sequence. In the meantime, they were organized to allow independent reading, according to the interest of the reader.

Chapter 2 of this book presents concepts, definitions, and procedures of network analysis involved in the real-time operation of EPSs. The chapter describes all stages of the PSSE process, especially Step 1, since the others will be dealt with in detail in the next chapters of this book. The chapter ends discussing the effect of TEs and PEs and the inclusion of Synchronized Phasor Measurements (SPMs) in the PSSE process. Finally, some practical problems are proposed.

Chapter 3 reviews the basic formulation of the power flow problem in EPS. This chapter also deals with the modeling of the electrical components of the network and the power flow equations that are also used in the formulation of the state estimation process. Finally, some theoretical and real-life application problems are proposed.

Chapter 4 deals with the third stage of the conventional PSSE process, that is, the state estimation itself. After the presentation of basic definitions, the formulations of the WLS state estimator are presented in the coupled, decoupled, and linear versions. Next, it is introduced the hybrid WLS estimator, which is the WLS estimator that allows the simultaneous processing of SCADA measurements and SPMs, as well as estimators statistically more robust than the WLS estimator. At the end of this chapter are some problems to assist in understanding the operation of the WLS Estimator.

Chapter 5 focuses on the qualitative characteristics of measurement sets in the context of static state estimation in EPSs, which refer to the analyses of observability and redundancy of measurements. Concepts, definitions, and some methods already developed to perform these analyses are presented in a detailed way considering the single-phase modeling of the power grid, the existence of only SCADA measurements. This is because most of the estimators and methods developed for observability and redundancy analysis work on these considerations. However, the methods that will be presented in this chapter for analysis of observability and redundancy of measurements can be extended to consider the three-phase modeling of the network as well as the existence of SCADA measurements and SPM, as will be discussed in the last sections of this chapter. At the end of this chapter are some problems to assist in understanding aspects related to the qualitative characteristics of measurement sets in the context of static state estimation in EPSs.

Chapter 6 covers the fourth step of the conventional PSSE process, the GE processing step, which is performed after processing the first three steps of the PSSE process, as presented in Chapter 2. This chapter presents definitions and the necessary theoretical basis for the understanding of the classical methodologies developed for GE processing from the WLS estimator residuals analysis. After presenting these methodologies, the chapter also provides some examples and exercises to enable a better understanding for GEs detection, identification, and elimination from the WLS estimator residuals analysis.

Chapter 7 describes the fundamentals of the innovation methodology, as well as its application in the processing of errors in the PSSE process. The chapter also provides some problems to enable a better understanding of the information presented.

Chapter 8 addresses the dynamic state estimation (DSE) problem, based on the Kalman filter (KF) formulation. The chapter presents the asymmetry index for anomaly detection and identification (in the context of DSE, suddenly large load changes as well as measurement GEs are named as anomaly). The chapter presents further several simulation results validating the presented anomaly detection model.

Chapter 9 presents the core ideas of data-driven state estimator solutions. Linear regression models are introduced, as well as machine-learning approaches used to estimate the gross errors on measurements. Hybrid data-driven physics-model-based solutions are further discussed. Problems concerning real-life applications are introduced.

The book also has two appendices, one with the definitions of statistical robustness and breakdown point and the other with answers to the exercises proposed in the book chapters.

References

Abur, A., Expósito, A., 2004. Power Systems State Estimation: Theory and Implementation. Marcel Dekker Inc., New York, USA.

Bretas, N.G., Piereti, S.A.R., Bretas, A.S., Martins, A.C.P., 2013. A geometrical view for multiple gross errors detection, identification, and correction in power system state estimation. IEEE Trans. Power Syst. 28 (3), 2128–2135.

Coutto Filho, M., Souza, J., 2009. Forecasting-aided state estimation part I: panorama. IEEE Trans. Power Syst. 24 (4), 1667–1677.

Coutto Filho, M., Silva, A., Falcão, D., 1990. Bibliography on power system state estimation (1968-1989). IEEE Trans. Power Syst. 5 (3), 950–961.

Falcão, D., Arias, M., 1994. State estimation and observability analysis based on echelon forms of the linearized measurement models. IEEE Trans. Power Syst. 9 (2), 979–987.

Falcão, D., Assis, S., 1988. Linear programming state estimation: error analysis and gross error identification. IEEE Trans. Power Syst. 3 (3), 809–815.

Irving, M., Owen, R., Sterling, M., 1978. Power-system state estimation using linear programming. Proc. Inst. Electr. Eng. 125 (9), 879–885.

Kotiuga, W.W., Vidyasagar, M., 1982. Bad data rejection properties of weighted least absolute value techniques applied to static state estimation. IEEE Trans. Power App. Syst. PAS-101 (4), 844–853.

Liu, Y., Ning, P., Reiter, M.K., 2009. False data injection attacks against state estimation in electric power grids. ACM Trans. Inf. Syst. Secur. 14 (1), 1–33.

Mili, L., Phaniraj, V., Rousseeuw, P., 1991. Least median of squares estimation in power systems. IEEE Trans. Power Syst. 6 (2), 511–523.

Monticelli, A., 1999. State Estimation in Electric Power Systems: A Generalized Approach. Kluwer Academic Publishers, Boston, USA.

Monticelli, A., 2000. Electric power system state estimation. Proc. IEEE 88 (2), 262–282.

Schweppe, F., 1970. Power system static-state estimation, part III: implementation. IEEE Trans. Power App. Syst. PAS-89 (1), 130–135.

Schweppe, F., Handschin, E., 1974. Static state estimation in electric power systems. Proc. IEEE 62 (7), 972–982.

Schweppe, F., Rom, D., 1970. Power system static-state estimation, part II: approximate model. IEEE Trans. Power App. Syst. PAS-89 (1), 125–130.

Schweppe, F., Wildes, J., 1970. Power system static-state estimation, part I: exact model. IEEE Trans. Power App. Syst. PAS-89 (1), 120–125.

Shivakumar, N., Jain, A., 2008. A review of power system dynamic state estimation techniques. In: POWERCON—IEEE Power India Conference, New Delhi, India, pp. 1–6.

Van Cutsem, T., Ribbens-Pavella, M., Mili, L., 1984. Hypothesis testing identification: a new method for bad data analysis in power system state estimation. IEEE Trans. Power App. Syst. PAS-103 (11), 3239–3252.

Wu, F., Moslehi, K., Bose, A., 2005. Power system control centers: past, present and future. Proc. IEEE 93 (11), 1890–1908.

Zhao, J., et al., 2019. Power system dynamic state estimation: motivations, definitions, methodologies, and future work. IEEE Trans. Power Syst. 34 (4), 3188–3198.

CHAPTER 2

Real-time operation of power systems

2.1 Operating states of an EPS

As discussed in Chapter 1, the first step in performing the EPS real-time operation is to determine the current operating state of the system, i.e., whether the system is operating adequately and, if it is not, it should indicate what must be done to correct this inadequate operation, that is, the appropriate control actions.

The operation of an EPS is subject to certain conditions which may be expressed in the form of three sets of constraints that determine the operating states of the EPS (Dy-Liacco, 1974; Monticelli, 1983):

Load Constraints—These constraints represent the fact that the power system must meet the load demand, i.e., it must supply power to all consumers. Therefore those are equality constraints expressed mathematically as:

$$g(x) = 0, \tag{2.1}$$

where x is the vector of state variables and g is a vector of nonlinear functions. Eq. (2.1) corresponds, in fact, to the imposition of Kirchhoff's First Law.

Operating Constraints—Operating constraints require that the operating limits of the system equipment (transmission lines, transformers, generators, etc.) must be respected, such as the maximum power that can be transmitted by a given transmission line. Thus inequality constraints are represented mathematically by:

$$h(x) \leq 0, \tag{2.2}$$

where h is a vector of nonlinear functions.

Security Constraints—In power systems, this characterization requires drawing up a list of possible (or likely) contingencies[a] in the current

[a] Such contingencies might include losses of power lines, losses of generating units, etc.

Cyber-Physical Power Systems State Estimation
https://doi.org/10.1016/B978-0-323-90033-1.00005-6

operating condition. Having this list of potential contingencies in hand, one can verify whether the system resists the impact of each of them without entering an inadequate state of operation, i.e., violating load and/or operation constraints. If s is a function vector consisting of all operation and load constraints for each contingency in the list, it is possible to characterize security constraints mathematically as:

$$s(x) \leq 0, \tag{2.3}$$

It is important to note that these constraints are vector functions of the state variables "x", which are the complex bus voltages of the system. Based on these constraints, four operating states can be defined (Monticelli, 1983, 1999):

1. Normal Secure State—It is the ideal operating state because the three sets of constraints are met: load, operation, and security. This means that the system is in perfect operating condition, with full supply of load demand and no violation of operating limits. In addition, there are compliance with security restrictions, i.e., none of the preestablished security contingencies, if they happen, will lead the system to an emergency. However, the occurrence of an unlisted contingency may cause an emergency.
2. Normal Alert State—Only load and operation restrictions are met in this state; however, not all security constraints are satisfied. In the same way as in the secure state, the system supplies all loads and without any violation of operating limits. Failure to observe security restrictions means that the occurrence of at least one of the contingencies listed as possible may lead the system to an emergency (violation of operating restrictions).
3. State of Emergency—This state is characterized by the violation of at least one operating restriction. The emergency can be caused by a contingency and consequent shutdown of one or more components of the system (lines, generators, transformers, etc.). All loads are supplied. Emergency elimination can be done by transitioning to the alert state, or by shutting down parts of the system (loads, for example), which brings the system to the next possible state of operation, the restorative.
4. Restorative State—Operating constraints are satisfied, but the system is not intact (violation of load restrictions, islands,[b] etc.). This state is

[b] Islanding can be succinctly defined as the physical separation of an area to the remainder of the connected system.

reached when an emergency is cleared by manual or automatic shutdown of parts of the system by the control center or by local devices. In the need to move from the state of emergency to the restorative, the integrity of the system is sacrificed in order to comply with the operating restrictions.

It can be concluded that the purpose of the real-time operation is to keep the system operating in the normal safe state of operation, in order to minimize transitions from that state to the emergency or restorative state.

Considering the operating states presented, the transitions between them can occur because of system disturbances (called involuntary transitions) or due to control actions (voluntary transitions) of the control and control centers. As shown in Monticelli (1983, 1999), the possible state transitions are:

- Secure-Alert Transition (Involuntary)—This type of transition may be caused by some unforeseen contingency, or even by the normal increase in load demand, which may take the system from secure state to alert state, in which one or more security constraints are no longer met.
- Secure-Alert Transition (Voluntary)—This transition results from the control center through the execution of the security control function (corrective mode), which is responsible for determining a viable solution for load, operation, and safety restrictions based on a case in which only security restrictions are violated. For this purpose, analyses are carried out in the EPS through, for example, an optimum power flow program with security restrictions, which allows the determination of a viable solution that minimizes the change of the system operating point, of control variables or the operating cost.
- Secure-Secure Transition (Voluntary)—Through data obtained by a short-term load forecasting program (1 h, for example), the operating point of the system is changed to fit the future conditions of the EPS, even though it is in the normal-secure state. This is because such future conditions could lead the system to alert or even, in extreme cases, to a state of emergency. In this case, the security control function can be activated in preventive mode, which will try to put the system in the secure state also for the expected conditions.
- Alert-Emergency Transition (Involuntary)—If the system is in the alert state and one of the previously listed contingencies responsible for that state occurs (or even in the occurrence of an unforeseen contingency), then the system will go into a state of emergency.

The state of emergency is reached when one of these contingencies occurs prior to adopting control measures to put the system in the secure state (alert-secure transition). If the system is in a state of emergency, two types of emergency control can operate: corrective or crisis.

- Emergency-Alert Transition (Voluntary)—This transition is performed by emergency control in the corrective mode. In this case, a constraint violation can be eliminated by control actions involving redispatching of the generation, variation of controlled voltages or transformer taps, switching capacitor banks or inductors, etc. In this mode of operation of the emergency control, contrary to what occurs in the crisis mode, the demand or the integrity of the system is not sacrificed. The emergency control actions in corrective mode can be preestablished, determined with the aid of computer programs during emergency or left to the operator, in the simplest cases.
- Emergency-Restorative Transition (Voluntary)—In some emergency situations, local control and protection devices act before any control center measure, placing the system in the restorative state by disconnecting loads or circuits. In other cases, these shutdowns are performed by the control center itself through the emergency control function in crisis mode, sacrificing the supply of loads. Such an action must be taken when it is not possible to meet the limits of operation by means of smoother means such as redispatching of generation or voltage control.
- Alert-Alert Transition (Voluntary)—This transition occurs when the load forecast of a system on alert indicates a possible emergency for a future state. For this reason, the operator can use the emergency control function in the preventive mode, which is responsible for changing the operating point of the system so that the previously detected emergency does not occur in the expected state.
- Restorative-Alert Transition (Voluntary)—This transition is done by performing the restorative control function, whose purpose is to reconnect the system loads and circuits, disconnected during the emergency, trying to bring the system back to the alert state or, if possible, to the secure state.

Fig. 2.1 represents a diagram of possible transitions between the various states of operation of an EPS.

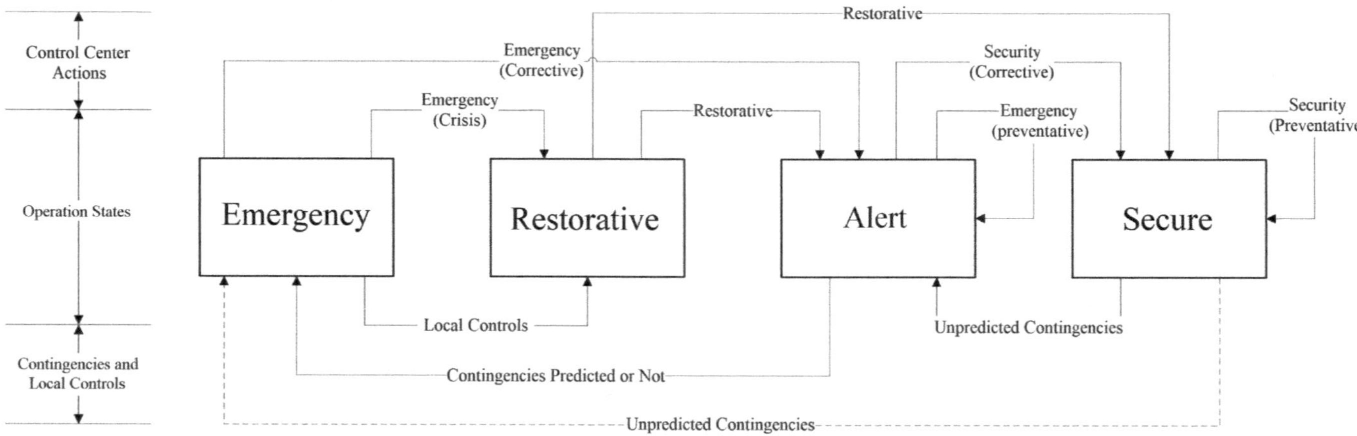

Fig. 2.1 Transition diagram of operating states of a power system.

2.2 Network analysis system

From the concepts previously presented, it can be said that the security of the system is the ability of an EPS, in the normal state, to undergo a disturbance without going into a state of emergency. Thus the safety of an EPS is directly related to the strategies adopted by the control centers, when the EPS is in the normal state of operation. By virtue of this, security strategies are divided into (Dy-Liacco, 1974):

- Security Monitoring: This step consists in the real-time identification of the topology and the current operating conditions of the system and possible violations of operating restrictions. For this, the data obtained in real time are processed in short time intervals.
- Security Analysis: Characterized by determining the security of the system against a set of contingencies previously established. It is performed through static and/or dynamic simulation programs, where the new operating state of the system is verified in the face of various contingencies.
- Preventive Control: If the security analysis implies that the system is in the normal alert state, the next step is to determine the actions that must be taken to bring the system into secure state if possible. Thus we aim to find the optimum solution that does not violate the operating, load, and security limits.

For the real-time execution of the aforementioned security strategies, the supervisory and control centers consist of several partial systems, among which is the network analysis system, which is responsible for performing functions such as the PSSE process, power flow, security analysis, corrective dispatching analysis, etc. However, due to the areas under authority of the companies and aiming to reduce the dimensions of the network analysis problems, the electric network is commonly divided into internal system and external system. The internal covers all buses and circuits owned and supervised by a given company, as well as supervised boundary areas. Differently, the external comprises all the unsupervised areas that are interconnected to the internal system. In this section, we will present the various functions that make up the network analysis system, in order to clarify its role in the real-time operation, as well as its interfaces.

1. Prefiltration Program: Performs compatibility tests on analog measurements in order to detect and exclude obvious errors, to differentiate them

from GEs, which are not obvious and require a treatment to be detected. Obvious errors can be detected by a simple plausibility check, even before the PSSE process. Some examples of errors such as these are negative voltage magnitudes, measurements with several orders of magnitude greater or less than the expected value, or measurements with large differences between input and output currents in a node of a substation.

2. The PSSE Process: Consists of a set of computational programs that allow the determination of the current operating state of the EPS. In order to do this, the input data of the PSSE process are representation of the electrical network at the bus section level, i.e., the representation of the physical connection of the system elements (generators, transformers, transmission lines, switches, breakers, etc.); continuous analog and prefiltered logic measurements; and the electrical parameters of the network. Traditionally the processing of this information takes place in four distinct stages that are executed sequentially:

   Step 1: Obtain the system topology with the corresponding configuration of meters in the bus-branch model.
   Step 2: Observability analysis and restoration.
   Step 3: State estimation.
   Step 4: Processing of GEs.

   Each of these steps will be presented in detail in the next section. It should be noted that if any GEs is identified in Step 4, a further running of Step 3 will be required. These two steps, third and fourth, will be performed iteratively until no GE is detected in Step 4, thus closing the PSSE process.

3. Load Forecasting Program: It is based on the demand forecast of the supervised and unsupervised network. This program is used by automatic generation control, which is responsible for power flow exchanges and frequency control, to better balance the power generated with the load demand and, consequently, to avoid abrupt changes in the frequency of the interconnected system. The load forecasting program also enables forecasting of the demand for the interconnected system by means of the distribution factors obtained by the analysis of consumption curves (daily or weekly).

4. Power Flow: Aims to determine the complex voltages (magnitudes and angles) of all the buses of the internal and external network of the EPS. For this, the network configurator (program responsible for executing

PSSE process Step 1) must previously contain information about the topology and the load forecast of the external network. In this program, the complex voltages are obtained by the execution of the power flow algorithm, considering the internal network buses as θV buses (from the results of the PSSE process), and the external network of the PV or PQ types.[c] It can also be used to verify the effects of actions of preventive or corrective controls, even before they are executed.
5. Security Analysis Program: Consists of subprograms responsible for the selection and simulation of the most plausible contingencies to occur in the EPS, in relation to the operating point of the system obtained through state estimation and power flow. These simulations can be static or dynamic. Dynamic simulations are still great challenges for the control centers, due to the great computational effort required and the processing time due to their complexity. However, static simulations, despite several approximations in the mathematical modeling of the EPS, give a good indication of the behavior of the system, and in addition, its computational requirements enable its use in real time. From the results of these simulations, we obtain the operating state resulting from a certain contingency.
6. Optimum Power Flow: It is commonly used to determine which are the best corrective control strategies that must be taken by the operation and its consequences for the system, due to some violation of operating limits. In this application, it is sought to find an optimal solution that satisfies the restrictions of operation, loads, and security, without changing excessively the point of operation of the system. To it, economic factors can be added, in order to reduce operating costs.

Fig. 2.2 shows a block representation of the network analysis system in the real-time context of EPS.

2.3 State estimation in electric power systems

In this section, four stages of the PSSE process will be presented, according to its classical formulation. That is, considering only SCADA measurements, the assumptions of balanced system and the configuration of the electrical network, and the parameters used in the models that represent it, are correct. Next, some aspects of the PSSE process not considered in the classical formulation will be addressed.

[c] Depending on which nodal variables are input as data and which are considered as unknowns, three types of buses are defined in Chapter 3.

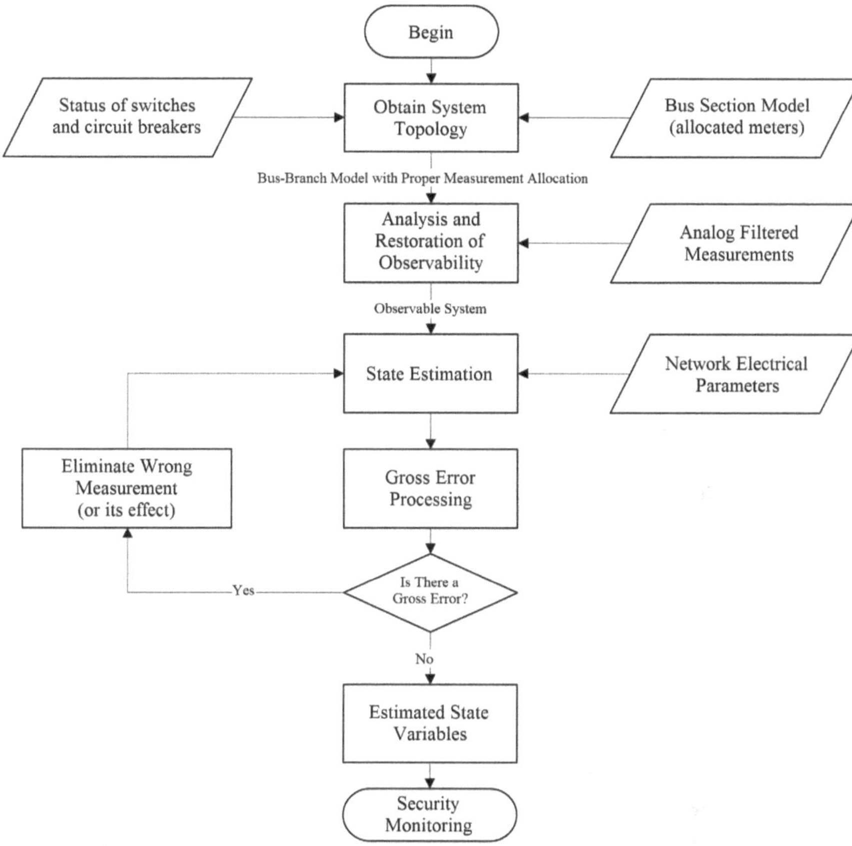

Fig. 2.2 Stages involved in the network analysis system in the context of EPS real-time operation.

2.3.1 Step 1: Obtaining the system topology with the corresponding meter configuration in the bus-branch model

This step is executed by the network configurator and consists in determining, in real-time, the current topology of the network and the corresponding configuration of meters in the bus-branch model. This is possible from the processing of data stored in a static database, which describes the connection of the system equipment (generators, current and potential transformers, loads, capacitors, transmission lines, meters, etc.) with the bus sections. In addition to these data, the configurator also processes the logic measurements, which are obtained continuously by the SCADA system and consist of statuses of switches and circuit breakers.

The process of obtaining the topology of the system by the network configurator has the following characteristics:

- Bus sections that are in the same voltage level can be interconnected by the closing of switching devices (switches and breakers) forming a single bus in the bus-branch model.
- For different combinations of status of the switching devices, we will, in general, have different network topologies.
- To obtain the initial topology of an EPS, the network configurator must process all the circuits (generators, transformers, transmission lines, etc.) and status of the available switching devices. After this initialization, the configurator will be executed again only in the event of changes in the position of the switching devices that cause changes in the initial topology of the EPS.

Example: Network configurator

To illustrate the process of obtaining the topology of the network in the bus-branch model with the corresponding configuration of meters, Figs. 2.3 and 2.4, respectively, present the information processed and the output of the network configurator.

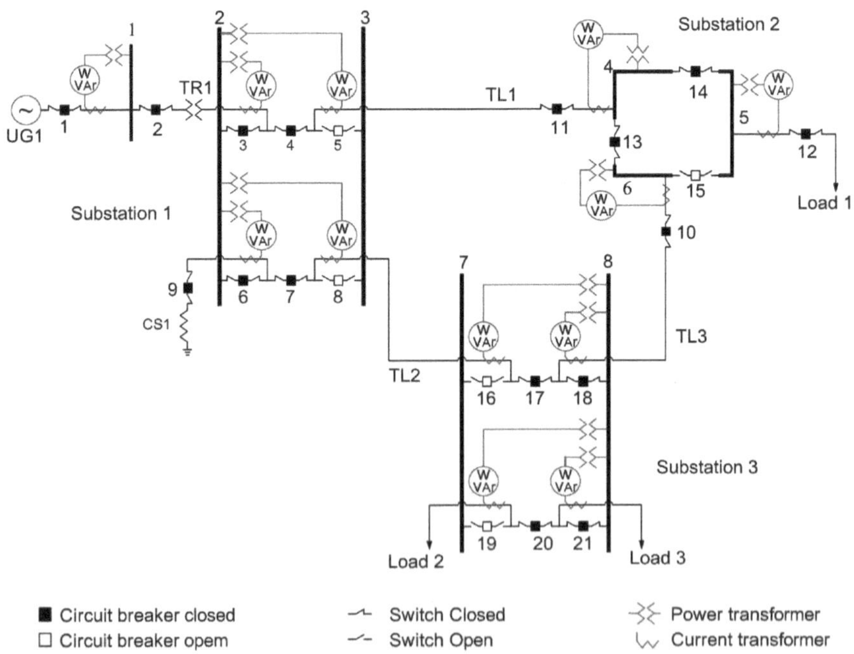

Fig. 2.3 Network model showing bus-section and switching devices (*GU*, generating unit; *SC*, synchronous condenser; *TL*, transmission line; *TR*, transformer; *VAr*, reactive power meter; *W*, active power meter) (Piereti et al., 2007).

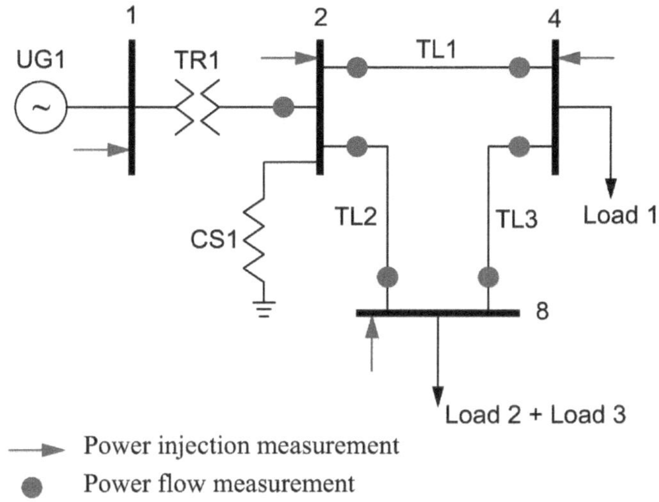

Fig. 2.4 EPS bus-branch model of Fig. 2.3, including meter location (Piereti et al., 2007).

As shown in Fig. 2.4, the configurator outputs the bus-branch model (with the corresponding meter association); the identification of each bus along with its generation, its loads and shunting devices; the connectivity between the buses due to the presence of transmission lines and transformers and, in addition, identifies islands and discards the islands that do not have generation (including buses and isolated branches). In general, the network configurators proposed for state estimation work in four distinct phases:

1. Substation configuration: In this phase, the information related to the bus sections of each substation is processed. All bus sections, connected by closed switching devices, will compose a network bus in the bus-branch model.
2. Network configuration: In this phase, the network buses determined in the previous phase are processed to verify that they are connected through closed switching devices to series-connected equipment (transmission lines and transformers). These connections will be represented as branches of the network, in the bus-branch model.
3. Association of available metering devices to the appropriate components in the bus-branch model: This phase is necessary because in practice the location of metering devices is given in terms of bus section and switching devices model. This phase is performed considering practical rules that depending on the meter arrangement may become a complex task (Monticelli, 1999).

4. Results tabulation: It consists of the tabulation of all the equipment connected to the network buses in the bus-branch model.

Due to the value of the process of obtaining system topology for network analysis, since 1973, several studies have been developed in the search for reliable and computationally efficient network configurators (Sasson et al., 1973; Dy-Liacco et al., 1973). In some situations, the change in the status of some logical device causes only changes in the generation, or the loading of certain buses, without changing the topology of the network. In such situations, since the matrices associated with the state estimation process do not change, a new execution of the network configurator is not necessary. This was a limitation of the first configurators developed, since even in the occurrence of a small change in the system, the entire network configuration process was started (Goderya et al., 1980; Bertran and Corbella, 1982).

In order to overcome this limitation (Prais and Bose, 1988), the concept of tracking configurator was introduced, which allows a quick updating of the network topology in the occurrence of small changes through a localized search by storing the configuration from the last execution.

Later, in Piereti et al. (2007) and Piereti et al. (2011), it was presented a network configurator that makes use of the data structure called Node-Depth Representation (Delbem et al., 2004). This configurator has the following interesting features:

- Allows the execution of the substation and network configuration phases at the same time, thus reducing the processing time required to obtain the bus-branch model;
- Enables the direct association of the meters to the network components in the bus-branch model;
- Enables a quick update of the network topology after status changes of switching devices, without the need to restart the entire network configuration process;
- Following changes in the status of switching devices, it indicates the necessary changes in the Jacobian matrix of the weighted least squares state estimator. From this information, partial matrix refactorization methods (Chan and Brandwajn, 1986) can be used to make the state estimation process faster.

2.3.2 Step 2: Observability analysis and restoration

The purpose of this step is to verify if the available analog and virtual measurements allow the estimation of state variables throughout the EPS (usually these variables are the complex nodal voltages). If so, the system is said to be observable. Otherwise, this lack of measurements can be solved in some situations by pseudo-measurements,[d] through which the system becomes observable. Another alternative for this situation is to determine the observable parts of the system, i.e., its observable islands.

In Clements and Wollember (1975), the minimum conditions for observability of an EPS were investigated, using its topology and the Kirchhoff laws. The concept of observable islands was then defined, and a heuristic algorithm was proposed. Although this algorithm was not capable of running in real time, it provided a basis for the concepts of algebraic, topological, and numerical observability presented in Krumpholz et al. (1980) and defined according to the formulation of the WLS state estimator.

Given the importance and the number of papers developed on the problem of observability analysis and restoration in the context of PSSE, Chapter 5 of this book will deal with this topic in detail.

2.3.3 Step 3: State estimation

This step aims to determine, in real-time, the state variables of the EPS, from the following information: the system topology and the corresponding configuration of meters in the bus-branch model, obtained by the network configurator in Step 1; the EPS parameters stored in the database; and the available set of analog and virtual measurements (in addition to pseudo-measurements that have been selected in Step 2 to restore observability).

To accomplish this task, several state estimators have been developed, which should allow the calculation of unknown state variables through a set of inexact analog measurements. Therefore the estimated values for these unknown state variables will also be inaccurate and the estimator must then find a way to obtain the best estimate for these state variables.

In view of the above, state estimators are based on statistical minimization criteria, among which, as reported in Chapter 1, the most used in EPS is the Weighted Least Squares (WLS). The WLS estimator will be presented in detail in Chapter 4.

2.3.4 Step 4: Processing gross errors in analog measurements

This step is necessary because the analog measurements provided to the state estimator are subject to GEs (due to, for example, analog-to-digital

[d] Load forecast data, generation forecast, historical data, etc. that are available in databases of control centers.

conversion errors, errors in telemetering communication channels, lack of measurement synchronism, saturation of measurement transformers, etc.), which take the state estimation process to state variables that do not correctly represent the current operating state of the EPS.

Traditionally, the first phase of this step is to detect the presence of GEs in the set of analog measurements processed in Step 3 by the state estimator. If no GE is detected, the PSSE process is terminated, and the state variables estimated in Step 3 will be used to define the current operating condition of the EPS. Otherwise, the next phase of Step 4 is the identification of the GE-carrying measurements. Then, the third and last phase of Step 4 is to eliminate the GE-carrying measurements, or at least the effects of these measurements on the estimated state variables. In this case, after the detection, identification, and elimination of GE-carrying measurements, Step 3 is performed again to obtain state variables without the influence of erroneous measurements.

The PSSE process ends when, after an estimate (Step 3), no GE in measurements is detected (Step 4).

Considering that the most widely used state estimator in the PSSE process is the WLS estimator, the most used methods for processing GEs are based on the statistical analysis of WLS estimator residuals. Some of these methods will be presented in Chapter 6.

2.3.5 Influence of topological and parameter errors in the PSSE process

As mentioned previously, the four steps of the traditional PSSE process were developed considering the assumptions that the network configuration and the parameters used in the models representing the elements of the electrical network are correct. Thus in the classical formulation of the PSSE process, only the possibility of GEs is considered, neglecting the possibility of occurrence of TEs and PEs. However, because those assumptions are often not true, TEs and PEs occur influencing the entire PSSE process, as will be discussed below.

Considering that Steps 2–4 of the PSSE process are based on the topology obtained in Step 1, if some TE occurs that has not been detected by the network configurator, such an error can cause an increase in the residuals of the analog measurements located in the vicinity of the wrongly configured elements of the system. Thus in Step 4, analog measurements with high normalized residuals are identified as GE carriers. In this situation, a process of eliminating analog measurements will be initiated, and the process may eventually reduce the level of local redundancy to zero. Therefore GE will no longer be detected in the analog measurement, but the TE remains. Consequently, the EPS model will not correctly represent its current operating condition. Using the errors analysis that will be presented in Chapter 7, since the measurement with error will be corrected, the last limitation will not occur.

Similar analysis can be performed considering PEs, since Steps 3 and 4 are based on the parameters provided to the estimator in Step 3. Thus if the information of some parameter of the system has been erroneously supplied to the state estimator, such an error will cause an increase in the residuals of the analog measurements, located in the vicinity of the element, whose parameter is the one that provided the wrong information.

Several methods have been developed to deal with PEs and TEs, mainly after the works proposing the representation of breakers and switches as zero impedance branches (Monticelli and Garcia, 1991; Monticelli, 1993a,b) in what is called generalized state estimation (Alsaç et al., 1998).

The generalized state estimation approach formulates and analyzes the analog, and switch and impedance data as a single interacting information set. In consequence, some parts of the network become modeled at the physical level (i.e., modeled in terms of bus-sections and switching devices) (Monticelli, 1999).

Reference (Abur and Exposito, 2004) presents a complete study of the influence of PEs and TEs on state estimation results, in which the most important factors and developed methods to deal with these errors are analyzed.

2.3.6 Synchronized phasor measurements in the PSSE process

In the attempt to improve the reliability of the PSSE process, which depends on the quality and redundancy of the available measurements, several surveys have been carried out since the mid-1980s on the possibility of using Synchronized Phasor Measurements (SPMs) (Thorp et al., 1985; Phadke et al., 1986; Hosseini et al., 2017).

The interest for the incorporation of SPMs in the PSSE process is justified mainly for the following reasons: the possibility of increasing the redundancy of the measurements; the fact that PMUs make measurements at short time intervals (samples rate of milliseconds) significantly faster than the SCADA system (samples rates of seconds); and the possibility of measuring the phase angles of the nodal voltages and branch currents of the electric network, variables whose measurement is impracticable through conventional SCADA measurements.

Basically, the structure of a synchronized phasor measurement system consists of the following elements: Phasor Measurement Units (PMUs), the basic element of the entire system; stations of reception of the clock signal of the Global Positioning System (GPS) (time base used for synchronization of the SPMs); and data concentrator and communication channels.

The PMU is a measurement instrument that was developed in the mid-1980s at the Virginia Polytechnic Institute and University in the United States and performs the synchronized sampling of the voltages and currents in the system, processes the sampled data by measuring the phasor quantities, and sends it to the data concentrator. The main purpose of the data concentrator is to receive and organize the phasor measurements sent by the PMUs, making them available for use in various applications, such as the PSSE process. Communication channels, for use in synchronized phasor measurement systems, include different structures, which operate either by cable (telephone lines, fiber-optic cables, or even the low and medium voltage network) or by wireless technology (Moraes, 2009).

It should be noted that systems formed only by conventional SCADA meters are based on steady-state analysis, not allowing to observe their dynamic characteristics. In systems containing PMUs, time-synchronized phasors are provided, usually 20, 30, or 60 samples per second, allowing the analysis of the dynamic behavior of the system.

Another advantage of using PMUs in the PSSE process is the possibility of making this process linear, if the system is completely observable by only SPMs (Phadke, 2002). This is because SPMs directly relate state variables (complex nodal voltages), unlike conventional measures of power, which relate measures and state variables through nonlinear functions.

However, due to the high costs involved in installing PMUs, mainly due to the need for telecommunication infrastructure, the current metering systems, destined for the PSSE process, will not be completely replaced by PMUs in a short period. This is the reason why state estimators including both SFMs and conventional measurements (power and voltage magnitude measurements obtained by the SCADA system) have been developed.

Hybrid state estimators able to deal simultaneously with both types of measurements have been proposed (Zhu and Abur, 2007; Korres and Manousakis, 2011). Although they present good performance with steady-state condition, the fact that PMUs provide information in a different time scale than the SCADA systems reduces the accuracy of those estimators, especially when changes occur on the network (Costa et al., 2013; Zhao et al., 2016; Massignan et al., 2019).

To overcome the problem of mixing in a single time scale measurements gathered at different instants, the forecasting-aided state estimation (FASE), a kind of dynamic state estimation method, established the ground base for modeling the temporal aspects of measurements and state variables (Coutto Filho and Souza, 2009; Zhao et al., 2019). In this context, some authors use Kalman filter-based estimators for integrating PMU measurements together with SCADA measurements. Among them, the extended Kalman filter is used in Netto et al. (2016) and the unscented Kalman filter in Wang et al. (2012). A weakness of these approaches is that they rely on the accuracy of the transition matrix in the forecasting stage, which is largely affected on the occurrence of system abrupt changes, such as load switching and contingencies (Zhao et al,. 2019; Massignan et al., 2020).

Another way for integrating fast sampling measurements in the PSSE process is the data fusion theory as presented in Costa et al. (2013). The proposed multiple stages state estimator processes each measurement type in two independent estimation stages and uses principles from data fusion to combine SCADA- and PMU-based estimates in an additional stage. Based on a Bayesian inference approach, a two-stage method able to perform information fusion from different kinds of sources was proposed in Massignan et al. (2020). The first stage of this method consists of the traditional WLS to deal with the SCADA measurements and builds a prior evaluation for the state variables. The second stage consists of using the PMU measurements to update the posterior distribution obtained according to Bayesian Inference concepts. Simulations with IEEE transmission systems show that the Bayesian two-stage method can provide accurate estimates, even for buses not observed by PMUs and when load variations occur (Massignan et al., 2020).

In Chapter 4, the formulations of both the classical WLS estimator (that considers only SCADA measurements) and the hybrid WLS estimator (that process simultaneously SCADA and SFMs) will be presented.

2.4 Exercise

Fig. 2.5 illustrates an EPS composed of three bus section-level substations, switching devices, and meters. Table 2.1 shows the legend used in the figure.

From the information presented in Fig. 2.5, the network topology with the corresponding meter association in the bus-branch model, shown in Fig. 2.6, is obtained by executing the network configurator.

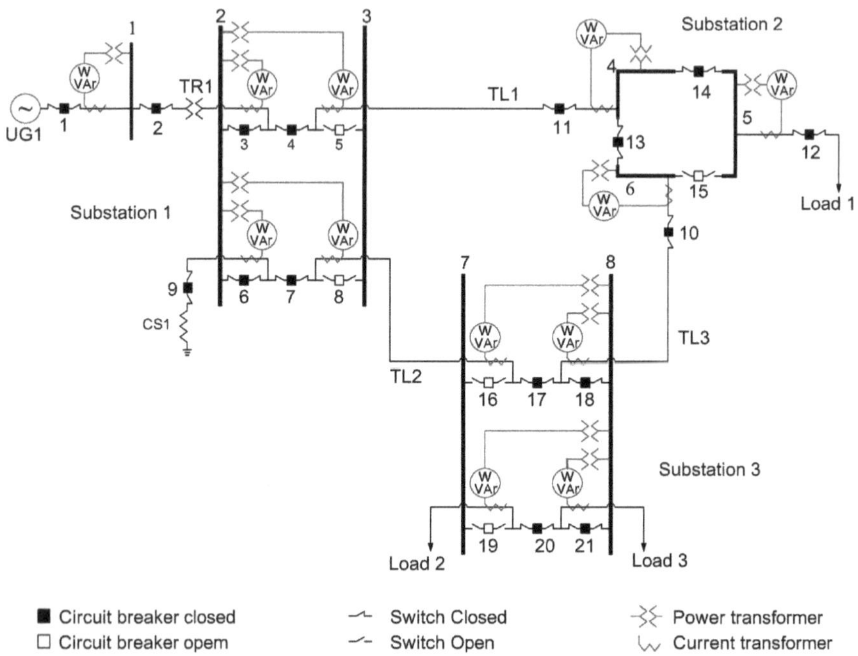

Fig. 2.5 EPS composed of three substations (*VAr*, reactive power meter; *W*, active power meter).

Table 2.1 Legend used for the system shown in Fig. 2.5.

UG	Generating unit
TR	Transformer
CS	Synchronous condenser
LT	Transmission line

Considering as a reference, the EPS shown in Fig. 2.6, with the respective status of switching devices, obtain the new topologies and the new meter configurations, in the bus-branch model, after performing each of the following switching operations:

(a) Opening the switching devices 3, 12, and 13, and closing the switching devices 5 and 8;

(b) Closing the switching device 16 and opening the switching device 18 (from the topology shown in Fig. 2.6);

(c) Closing the switching device 16 and opening the switching devices 4 and 10 (from the topology shown in Fig. 2.6).

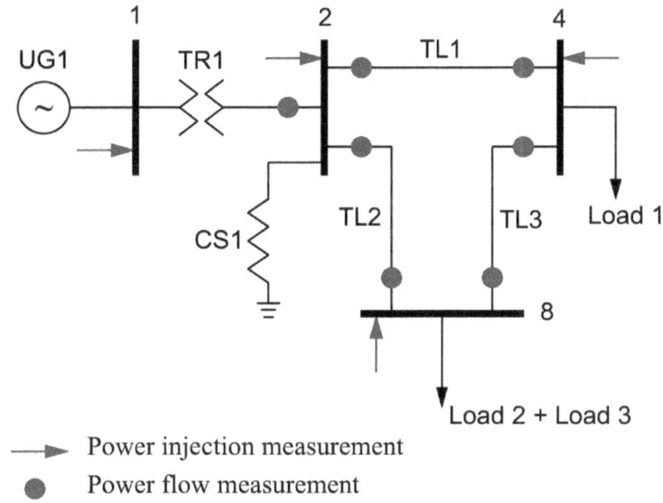

Fig. 2.6 Topology of the EPS presented in Fig. 2.5 in the bar-branch model, with the corresponding association of meters.

References

Abur, A., Exposito, A., 2004. Power Systems State Estimation: Theory and Implementation. Marcel Dekker, Inc., New York, USA.

Alsaç, O., Vempati, N., Monticelli, A., 1998. Generalized state estimation. IEEE Trans. Power Syst. 13 (3), 1069–1075.

Bertran, M., Corbella, X., 1982. On the validation and analysis of a new method for power network connectivity determination. IEEE Trans. Power App. Syst. PAS-101 (2), 316–324.

Chan, S., Brandwajn, V., 1986. Partial matrix refactorization. IEEE Power Eng. Rev. PER-6 (2), 44–45.

Clements, K., Wollember, B., 1975. An algorithm for observability determination in power system state estimation. In: 1975 IEEE PES Summer Meeting, San Francisco, USA, pp. 447–453.

Costa, A.S., Albuquerque, A., Bez, D., 2013. An estimation fusion method for including phasor measurements into power system real-time modeling. IEEE Trans. Power Syst. 28 (2), 1910–1920.

Coutto Filho, M., Souza, J., 2009. Forecasting-aided state estimation part I: panorama. IEEE Trans. Power Syst. 24 (4), 1667–1677.

Delbem, A., Carvalho, A., Bretas, N., 2004. Graph chain representation associated to an evolutionary algorithm for restoration of radial distribution systems. Int. J. Eng. Intell. Syst. Electr. Eng. Commun. 12, 3–12.

Dy-Liacco, T.E., 1974. Real-time computer control of power systems. Proc. IEEE 62 (7), 884–891.

Dy-Liacco, T.E., Ramarao, K., Weiner, A., 1973. Network status analysis for real-time systems. In: Power Industry Computer Applications Conference, Minneapolis, Minnesota, June, pp. 356–362.

Goderya, F., Metwally, A., Mansour, O., 1980. Fast detection and identification of islands in power networks. IEEE Trans. Power App. Syst. PAS-99 (1), 217–221.
Hosseini, S.A., Sadeh, J., Mozafari, B., 2017. Wide-area fault location method considering gross measurement errors. IET Gener. Transm. Distrib. 11 (18), 4670–4679.
Korres, G.N., Manousakis, N.M., 2011. State estimation and bad data processing for systems including PMU and SCADA measurements. Electr. Power Syst. Res. 81 (7), 1514–1524.
Krumpholz, G., Clements, K., Davis, P., 1980. Power system observability: a practical algorithm using network topology. IEEE Trans. Power App. Syst. PAS-99 (4), 1534–1542.
Massignan, J.A.D., London Jr., J.B.A., Maciel, C.D., Bessani, M., Miranda, V., 2019. PMUs and SCADA measurements in power system state estimation through Bayesian inference. In: 2019 IEEE Milan Powertech, Milan, Italy, pp. 1–6.
Massignan, J.A.D., London Jr., J.B.A., Miranda, V., 2020. Tracking power system state evolution with maximum-correntropy-based extended Kalman filter. J. Mod. Power Syst. Clean Energy 8 (4), 616–626.
Monticelli, A., 1983. Fluxo de Carga em Redes de Energia Elétrica. São Paulo, Brasil, Editora Edgard Blucher LTDA (in Portuguese).
Monticelli, A., 1993a. Modeling circuit breakers in weighted least squares state estimation. IEEE Trans. Power Syst. 8 (3), 1143–1149.
Monticelli, A., 1993b. The impact of modeling short circuit branches in state estimation. IEEE Trans. Power Syst. 8 (1), 364–370.
Monticelli, A., 1999. State Estimation in Electric Power Systems: A Generalized Approach. Kluwer Academic Publishers, Boston, EUA.
Monticelli, A., Garcia, A., 1991. Modeling zero impedance branches in power systems state estimation. Proc. IEEE/PES Winter Meeting 8 (1), 364–370.
Moraes, R., 2009. Sincrofasores em Sistemas de Potência: aplicações na estimação de estado (Ph.D. Thesis). Universidade Federal Fluminense, Rio de Janeiro, Brazil (in Portuguese).
Netto, M., Zhao, J., Mili, L., 2016. A robust extended Kalman filter for power system dynamic state estimation using PMU measurements. In: 2016 IEEE PES General Meeting, July, pp. 1–5.
Phadke, A.G., 2002. Synchronized phasor measurements-a historical overview. In: IEEE/PES Transmission and Distribution Conference and Exhibition, Yokohama, Japan, vol. 1, pp. 476–479.
Phadke, A.G., Thorp, J.S., Karimi, K.J., 1986. State estimation with phasor measurements. IEEE Trans. Power Syst. 1 (1), 233–238.
Piereti, S.A.R., Delbem, A.C.B., London Jr., J.B.A., Bretas, N.G., 2007. Tracking network topology processor using node-depth representation. In: 2007 IEEE Lausanne Power Tech, Lausanne, pp. 143–148.
Piereti, S., London Jr., J., Delbem, A., Bretas, N., 2011. Configurador de redes baseado na representação nó profundidade para efeito de estimação de estados: simulação com sistemas de 6 e 14 barras. In: IEEE Sinergia IFSP. 12, pp. 23–36.
Prais, M., Bose, A., 1988. A topology processor that tracks network modifications. IEEE Trans. Power Syst. 3 (3), 992–998.
Sasson, A., Ehrmann, S., Lynch, P., Van Slyck, L., 1973. Automatic power system network topology determination. IEEE Trans. Power App. Syst. PAS-92 (2), 610–618.
Thorp, J.S., Phadke, A.G., Karimi, K.J., 1985. Real time voltage-phasor measurement for static state estimation. IEEE Trans. Power App. Syst. PAS-104 (11), 3098–3106.
Wang, S., Gao, W., Meliopoulos, A.P.S., 2012. An alternative method for power system dynamic state estimation based on unscented transform. IEEE Trans. Power Syst. 27 (2), 942–950.

Zhao, J., Zhang, G., Das, K., Korres, G.N., Manousakis, N.M., Sinha, A.K., He, Z., 2016. Power system real-time monitoring by using PMU based robust state estimation method. IEEE Trans. Smart Grid 7 (1), 300–309.
Zhao, J., et al., 2019. Power system dynamic state estimation: motivations, definitions, methodologies, and future work. IEEE Trans. Power Syst. 34 (4), 3188–3198.
Zhu, J., Abur, A., 2007. Effect of phasor measurements on the choice of reference bus for state estimation. In: 2007 IEEE PES General Meeting, June, pp. 1–5.

CHAPTER 3

Power flow in electrical systems

3.1 Introduction

The problem of power flow, or load flow, in a power grid basically consists of determining the operating conditions (complex voltages in the buses and distribution of power flows) as a function of the network topology in the bus-branch model, as defined in Chapter 2, and the levels of load demand and power generation. It is a basic tool for the study of electrical networks, used both in the planning and operation of EPSs. In this problem, the system modeling is static, which entails in the representation of the electrical network through a set of equations and nonlinear algebraic inequalities corresponding, respectively, to the Kirchhoff laws and to the operational restrictions of the system. Such a representation is used in situations in which the variations of electrical quantities with time are slow enough that the transient effects can be ignored.

In general, in the problems of power flow and state estimation, the components of an electric energy system can be classified into two groups (Monticelli, 1999; Sttot, 1974):

- Those that are connected between any bus of the network and the bus that represents the ground node, as is the case of generators, loads, reactors, and capacitors.
- Those that are connected between any two buses of the network, as is the case of transmission lines and transformers.

The generators and loads are considered external parts of the system and are modeled by power injections in the network buses. The basic equations of power flow are obtained by imposing the conservation of the active and reactive powers in each bus of the network (Kirchhoff's First Law), that is, the net power injected must be equal to the sum of the powers that flow through the internal components that have this bus as one of their terminals.

3.2 Basic problem formulation

In the basic power flow formulation, four nodal variables are assigned to each bus of the network, and according to the type of bus, two are considered

Cyber-Physical Power Systems State Estimation
https://doi.org/10.1016/B978-0-323-90033-1.00008-1

Table 3.1 Power flow variables and each bus type.

Type	Data	Unknown	Characteristics
PQ	P_k, Q_k	V_k, θ_k	Represent loading buses
PV	P_k, V_k	Q_k, θ_k	Represent generation buses
$V\theta$	V_k, θ_k	P_k, Q_k	Generation buses (reference)

as input data and two as unknowns of the problem. For any bus k, these nodal variables are:

V_k voltage magnitude at bus k;
θ_k voltage phase angle at bus k;
P_k net active power (generation minus load) at bus k;
Q_k net reactive power at bus k.

Depending on which, nodal variables are input as data and which are considered as unknowns, three types of buses are defined, as shown in the following table (Table 3.1):

The $V\theta$[a] bus is generally assigned to a large capacity generating unit and has two functions (Sttot, 1974):

- Provide an angular reference to the network (the voltage magnitude reference is the ground node itself);
- Balance the power supply and demand of the system by considering the transmission losses not known before solving the problem. This function justifies the need of having a bus in the system to which the active power is not initially specified.

The set of equations is formed by two equations for each bus, each representing the fact that the active and reactive powers injected into a bus are equal to the sum of the corresponding flows that leave the bus through transmission lines, transformers, etc., as shown in Fig. 3.1. This is equivalent to the imposition of Kirchhoff's First Law.

For this representation, one has

$$\begin{aligned} P_k &= \sum_{l \in \Omega_k} P_{kl}(V_k, V_l, \theta_k, \theta_l) \\ Q_k + Q_k^{sh} &= \sum_{l \in \Omega_k} Q_{kl}(V_k, V_l, \theta_k, \theta_l) \end{aligned} \tag{3.1}$$

where $k = 1, \ldots, n$, where n is the number of buses in the network; Ω_k is the set of buses adjacent to bus k; V_k, V_l are the voltage magnitudes in

[a] Also known as reference, slack, or swing bus.

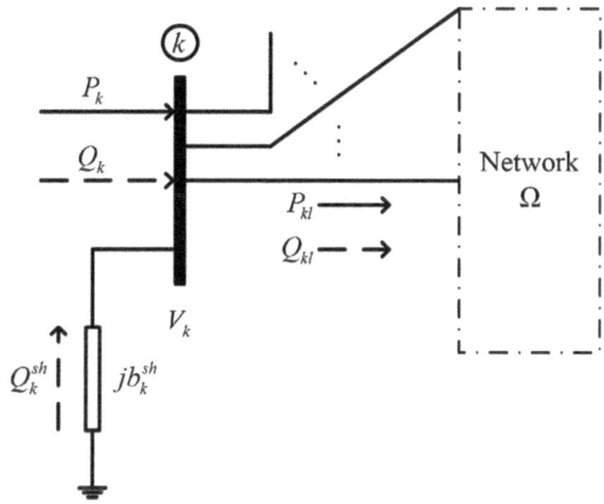

Fig. 3.1 Representation of Kirchhoff's First Law imposed on a bus of the system.

the terminal buses of the $k-l$ branch; θ_k, θ_l are the phase angles of the voltages in the terminal buses of the $k-l$ branch; P_{kl}, Q_{kl} are the active and reactive power flows, respectively, in the $k-l$ branch; Q_k^{sh} is the component of the reactive power injection due to the shunt element of the bus k ($Q_k^{sh}=b_k^{sh}\cdot V_k^2$, where b_k^{sh} is the shunt susceptance connected to the bus k).

The problem of power flow is undetermined for all variables θ; therefore it is necessary to adopt an angular reference given by the bus $V\theta$. The equations in (3.1) were written considering the following signal convention:

- Net power injections are positive when they enter the bus (generation) and negative when they leave the bus (load).
- Power flows are positive when they exit the bus and negative when they enter it.
- For the shunt elements of the buses, the same convention is adopted as for the injections.

The set of inequalities, which are also part of the basic formulation of the power flow problem, is formed by the constraints in the magnitudes of the nodal voltages in the PQ buses and by the limits of reactive power injections of the PV buses, as follows:

$$\begin{aligned} V_k^{\min} \leq V_k \leq V_k^{\max} \\ Q_k^{\min} \leq Q_k \leq Q_k^{\max} \end{aligned} \tag{3.2}$$

In addition to these restrictions in the basic formulation of the problem, it is also possible to include the limits of the tap values of the in-phase and phase-shifting transformers, the limits on the generation capacity of buses responsible for the tie-line control, the limits on the magnitudes of the voltages of the PV buses, etc. (Monticelli, 1999).

3.3 Modeling and equations for calculation of power flow

The flow and power injection equations necessary to study power flow and state estimation in EPS will be presented in the following.

3.3.1 Equation of power flows between buses

For the development of the expressions of power, flow between buses will be considered, initially, the models of transmission lines, in-phase transformers, and phase-shifting transformers. Next, it will be presented a general formulation, based on the generalized representation of the equivalent model of transmission lines and transformers, known as generalized π Model (Monticelli, 1999).

Transmission lines

The equivalent π model of a transmission line, represented in Fig. 3.2, is composed of three parameters (Monticelli, 1999): series resistance r_{kl}, series reactance x_{kl}, and shunt susceptance b_{kl}^{sh}.

Since the expression of the series impedance in terms of the parameters is given by

$$z_{kl} = r_{kl} + jx_{kl}, \tag{3.3}$$

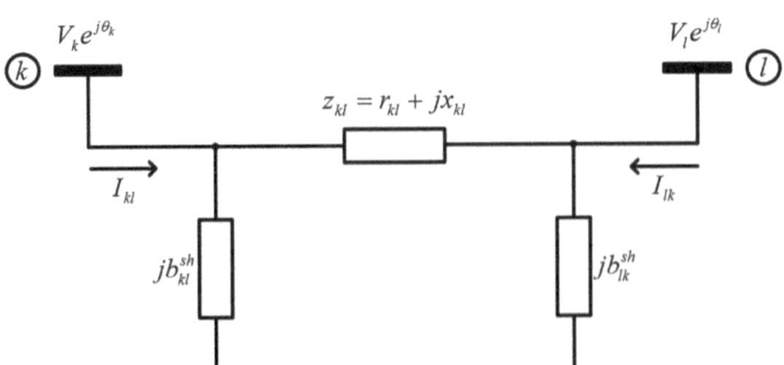

Fig. 3.2 Equivalent π model of a transmission line.

and the type of circuit analysis employed is the nodal, it becomes necessary to work with the series parameters in terms of line conductance and susceptance. Therefore the series admittance of the line is given by

$$y_{kl} = z_{kl}^{-1} = g_{kl} + jb_{kl}, \tag{3.4}$$

where the series conductance (g_{kl}) and the series susceptance (b_{kl}) are given by

$$\begin{aligned} g_{kl} &= \frac{r_{kl}}{r_{kl}^2 + x_{kl}^2} \\ b_{kl} &= -\frac{x_{kl}}{r_{kl}^2 + x_{kl}^2} \end{aligned} \tag{3.5}$$

Note 3.1: For the equivalent π model of a transmission line, r_{kl} and x_{kl} are positives, which imply positive g_{kl} and negative (or inductive) b_{kl}. The b_{kl}^{sh} element is positive, since the line shunt element is capacitive.

From the analysis of the current I_{kl}, as shown in Fig. 3.2, it is observed that it is composed by two components: one in series (branch of the impedance z_{kl}) and another shunt (branch of the susceptance b_{kl}^{sh}). Thus through the nodal analysis, we have the following expressions:

$$\begin{aligned} I_{kl} &= y_{kl}(E_k - E_l) + jb_{kl}^{sh} E_k \\ I_{lk} &= y_{kl}(E_l - E_k) + jb_{kl}^{sh} E_l \end{aligned} \tag{3.6}$$

where $E_k = V_k e^{j\theta_k}$ and $E_l = V_l e^{j\theta_l}$.

Based on the relationships of voltages and currents, we obtain the equation of the complex power flow through a transmission line:

$$\begin{aligned} S_{kl}^* &= P_{kl} - jQ_{kl} = E_k^* I_{kl} \\ S_{kl}^* &= E_k^* \left[y_{kl}(E_k - E_l) + jb_{kl}^{sh} E_k \right] \\ S_{kl}^* &= y_{kl} V_k e^{-j\theta_k} \left(V_k e^{j\theta_k} - V_l e^{j\theta_l} \right) + jb_{kl}^{sh} V_k^2 \\ S_{kl}^* &= y_{kl} V_k^2 - y_{kl} V_k V_l e^{-j(\theta_k - \theta_l)} + jb_{kl}^{sh} V_k^2 \end{aligned} \tag{3.7}$$

Considering $\theta_{kl} = \theta_k - \theta_l$, $e^{j\theta_{kl}} = \cos\theta_{kl} + j\sin\theta_{kl}$, $e^{-j\theta_{kl}} = \cos\theta_{kl} - j\sin\theta_{kl}$, and $y_{kl} = g_{kl} + jb_{kl}$, we have

$$S_{kl}^* = (g_{kl} + jb_{kl}) V_k^2 - (g_{kl} + jb_{kl}) V_k V_l (\cos\theta_{kl} - j\sin\theta_{kl}) + jb_{kl}^{sh} V_k^2 \tag{3.8}$$

By separating the real and imaginary parts of Eq. (3.8), we obtain the active and reactive flows:

$$\begin{aligned} P_{kl} &= V_k^2 g_{kl} - V_k V_l g_{kl} \cos\theta_{kl} - V_k V_l b_{kl} \sin\theta_{kl} \\ Q_{kl} &= -V_k^2 \left(b_{kl} + b_{kl}^{sh} \right) + V_k V_l b_{kl} \cos\theta_{kl} - V_k V_l g_{kl} \sin\theta_{kl} \end{aligned} \tag{3.9}$$

Similarly, and considering $\cos\theta_{lk} = \cos\theta_{kl}$ and $\sin\theta_{lk} = -\sin\theta_{kl}$, the equivalent P_{lk} and Q_{lk} flows are obtained:

$$\begin{aligned} P_{lk} &= V_l^2 g_{kl} - V_k V_l g_{kl} \cos\theta_{kl} + V_k V_l b_{kl} \sin\theta_{kl} \\ Q_{lk} &= -V_l^2 \left(b_{kl} + b_{kl}^{sh}\right) + V_k V_l b_{kl} \cos\theta_{kl} + V_k V_l g_{kl} \sin\theta_{kl} \end{aligned} \tag{3.10}$$

Transformers

The equivalent transformer model used in power flow and state estimation studies is composed of a series impedance z_{kl} and an ideal autotransformer in the primary, with a transformation ratio given by t_{kl}. Such a model can be seen in Fig. 3.3.

As can be seen in Fig. 3.3, p denotes a reference point for the transformation ratio, so the voltage magnitude at this point, in relation to bus k, is given by $V_p = a_{kl} V_k e^{j\phi}$, where φ is the value of the lag caused by the transformer.

Note 3.2: If the data of the transformer follows the representation $\overline{t_{kl}} : 1$, simply convert the relationship as follows:

$$t_{kl} = \frac{1}{\overline{t_{kl}}} \tag{3.11}$$

For example, if a 500/750 kV transformer has a tap-turns ratio of 1.050:1 on the low voltage side (500 kV) and without lag, then $t_{kl} = 1/1.050 = 0.9524$.

In-phase transformer

In general, the in-phase transformer modeling comprises a series impedance or admittance and an ideal autotransformer (no losses in the core) whose transformation ratio is given by $1 : a_{kl}$. Fig. 3.4 illustrates this type of transformer connecting buses k and l.

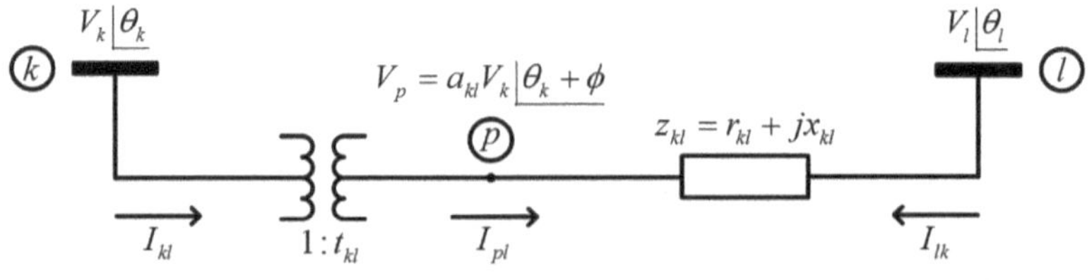

Fig. 3.3 Transformer model with complex transformation ratio $t_{kl} = a_{kl}\, e^{j\phi}$.

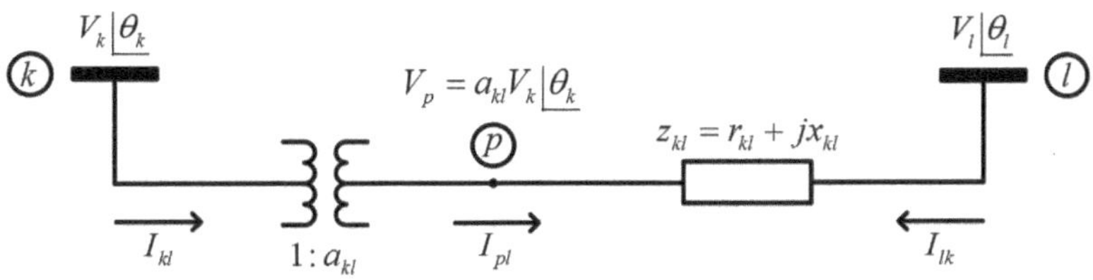

Fig. 3.4 Representation of an in-phase transformer.

As can be seen, p denotes a reference point for the transformation ratio. Thus the relationship between the magnitude of voltage at this point and bus k is given by a_{kl}, that is, $V_p = a_{kl}V_k$. Since in this case there is no angular lag between k and p ($\theta_k = \theta_p$), the relationship between the complex voltages is given by

$$\frac{E_p}{E_k} = \frac{V_p e^{j\theta_p}}{V_k e^{j\theta_k}} = a_{kl} \tag{3.12}$$

From the ideal model, that is, without considering the losses in the transformer, the following relationship is valid:

$$\begin{gathered} E_k I_{kl}^* + E_p I_{lk}^* = 0 \\ E_k I_{kl}^* = -a_{kl} E_k I_{lk}^* = 0 \end{gathered} \tag{3.13}$$

therefore $\frac{I_{kl}}{I_{lk}} = -\frac{|I_{kl}|}{|I_{lk}|} = -a_{kl}$.

By performing the nodal analysis of the transformer model, shown in Fig. 3.4, in terms of the complex currents I_{kl} and I_{lk}, the following equations are obtained:

$$\begin{gathered} I_{kl} = a_{kl} I_{pl} = a_{kl}(-I_{lk}) \\ I_{kl} = -a_{kl} y_{kl}\left(E_l - E_p\right) = -a_{kl} y_{kl} E_l + a_{kl} y_{kl} E_p \end{gathered} \tag{3.14}$$

as $E_p = E_k a_{kl}$, we have that:

$$I_{kl} = -a_{kl} y_{kl} E_l + a_{kl}^2 y_{kl} E_k \tag{3.15}$$

and

$$\begin{gathered} I_{lk} = y_{kl}\left(E_l - E_p\right) = y_{kl} E_l - y_{kl} E_p \\ I_{lk} = y_{kl} E_l - a_{kl} y_{kl} E_k \end{gathered} \tag{3.16}$$

Based on the relationships of voltages and currents, we must follow the equation of the complex power flow from bus k to bus l:

$$\begin{aligned}
S_{kl}^{*} &= P_{kl} - jQ_{kl} = E_{k}^{*} I_{kl} \\
S_{kl}^{*} &= E_{k}^{*}\left[-a_{kl}y_{kl}E_{l} + a_{kl}^{2}y_{kl}E_{k}\right] \\
S_{kl}^{*} &= V_{k}e^{-j\theta_{k}}\left[-a_{kl}y_{kl}V_{l}e^{j\theta_{l}} + a_{kl}^{2}y_{kl}V_{k}e^{j\theta_{k}}\right] \\
S_{kl}^{*} &= y_{kl}a_{kl}V_{k}e^{-j\theta_{k}}\left[-V_{l}e^{j\theta_{l}} + a_{kl}V_{k}e^{j\theta_{k}}\right] \\
S_{kl}^{*} &= -y_{kl}a_{kl}V_{k}V_{l}e^{-j(\theta_{k}-\theta_{l})} + y_{kl}a_{kl}^{2}V_{k}^{2}
\end{aligned} \tag{3.17}$$

Considering $\theta_{kl}=\theta_k-\theta_l$, $e^{-j\theta_{kl}}=\cos\theta_{kl}-j\sin\theta_{kl}$ and $y_{kl}=g_{kl}+jb_{kl}$, we have

$$S_{kl}^{*} = -(g_{kl}+jb_{kl})a_{kl}V_kV_l(\cos\Theta_{kl}-jsin\Theta_{kl}) + (g_{kl}+jb_{kl})a_{kl}^{2}V_k^{2} \tag{3.18}$$

Separating the real and imaginary part of Eq. (3.18), we obtain the equations for the active and reactive power flows:

$$\begin{aligned}
P_{kl} &= g_{kl}a_{kl}^{2}V_k^{2} - a_{kl}V_kV_lg_{kl}\cos\theta_{kl} - a_{kl}V_kV_lb_{kl}\sin\theta_{kl} \\
Q_{kl} &= -b_{kl}a_{kl}^{2}V_k^{2} + a_{kl}V_kV_lb_{kl}\cos\theta_{kl} - a_{kl}V_kV_lg_{kl}\sin\theta_{kl}
\end{aligned} \tag{3.19}$$

Following the same procedure, we have the equation of the power flow from bus l to bus k:

$$\begin{aligned}
S_{lk}^{*} &= P_{lk} - jQ_{lk} = E_{l}^{*} I_{lk} \\
S_{lk}^{*} &= E_{l}^{*}\left[y_{kl}E_{l} - a_{kl}y_{kl}E_{k}\right] \\
S_{lk}^{*} &= V_{l}e^{-j\theta_{l}}\left[y_{kl}V_{l}e^{j\theta_{l}} - a_{kl}y_{kl}V_{k}e^{j\theta_{k}}\right] \\
S_{lk}^{*} &= y_{kl}V_{l}^{2} - a_{kl}y_{kl}V_{k}V_{l}e^{-j(\theta_{l}-\theta_{k})}
\end{aligned} \tag{3.20}$$

If $\theta_{lk}=\theta_l-\theta_k$, $e^{-j\theta_{lk}}=\cos\theta_{lk}-j\sin\theta_{lk}$ and $y_{kl}=g_{kl}+jb_{kl}$, we have:

$$S_{lk}^{*} = (g_{kl}+jb_{kl})V_l^{2} - a_{kl}(g_{kl}+jb_{kl})V_kV_l(\cos\Theta_{lk}-jsin\Theta_{lk}) \tag{3.21}$$

Separating the real and imaginary part of Eq. (3.21), we obtain the equations for the active and reactive flows:

$$\begin{aligned}
P_{lk} &= g_{kl}V_l^{2} - a_{kl}V_kV_lg_{kl}\cos\theta_{lk} - a_{kl}V_kV_lb_{kl}\sin\theta_{lk} \\
Q_{lk} &= -b_{kl}V_l^{2} + a_{kl}V_kV_lb_{kl}\cos\theta_{lk} - a_{kl}V_kV_lg_{kl}\sin\theta_{lk}
\end{aligned} \tag{3.22}$$

Phase-shifting transformer

The phase-shifting transformers are devices capable of controlling the phase shift, or phase lag between the primary and secondary voltages, and thus

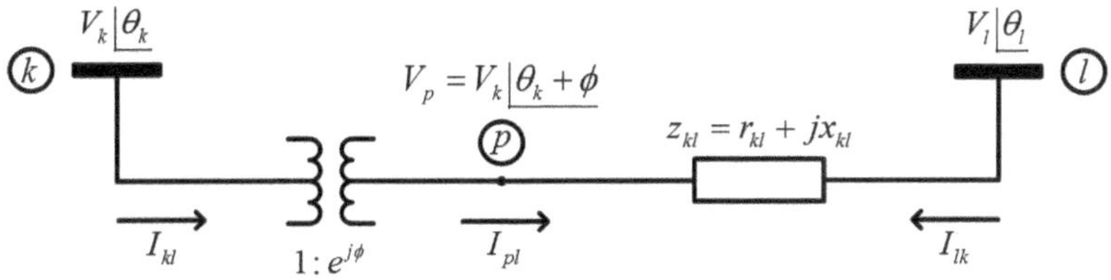

Fig. 3.5 Representation of a phase-shift transformer.

provide active power flow control between the buses. Fig. 3.5 illustrates this type of transformer connecting the buses k and l.p is a reference point for the transformation ratio, so the relationship between the complex voltage at this point by the bus k is given by $e^{j\phi}$, that is, $E_p = E_k e^{j\phi}$, where ϕ is the value of the phase difference caused by the transformer.

By the nodal analysis of the circuit, the expression of the complex current I_{kl} is given by

$$\begin{aligned} I_{kl} &= e^{-j\phi} I_{pl} = e^{-j\phi}(-I_{lk}) \\ I_{kl} &= -e^{-j\phi}\left(E_l - E_p\right) y_{kl} \\ I_{kl} &= -y_{kl} E_l e^{-j\phi} + y_{kl} E_p e^{-j\phi} \end{aligned} \tag{3.23}$$

as $E_p = E_k e^{j\phi}$, we have

$$I_{kl} = -y_{kl} E_l e^{-j\phi} + y_{kl} E_k \tag{3.24}$$

Analogously, we obtain the current I_{lk}:

$$\begin{aligned} I_{lk} &= \left(E_l - E_p\right) y_{kl} \\ I_{lk} &= y_{kl} E_l - y_{kl} E_k e^{j\phi} \end{aligned} \tag{3.25}$$

Based on the relationships of voltages and currents, we have the following equation derivation for the complex power flow from bus k to bus l:

$$\begin{aligned} S_{kl}^* &= P_{kl} - jQ_{kl} = E_k^* I_{kl} \\ S_{kl}^* &= E_k^*\left[-y_{kl} E_l e^{-j\phi} + y_{kl} E_k\right] \\ S_{kl}^* &= V_k e^{-j\theta_k}\left[-y_{kl} V_l e^{j\theta_l} e^{-j\phi} + y_{kl} V_k e^{j\theta_k}\right] \\ S_{kl}^* &= y_{kl} V_k V_l e^{-j(\theta_k - \theta_l + \phi)} + y_{kl} V_k^2 \end{aligned} \tag{3.26}$$

Considering $e^{-j(\theta_{kl}+\phi)} = \cos(\theta_{kl}+\phi) - j\sin(\theta_{kl}+\phi)$ and $y_{kl} = g_{kl} + jb_{kl}$, we have:

$$S_{kl}^* = (g_{kl} + jb_{kl}) V_k V_l[\cos(\theta_{kl} + \phi) + jsin(\theta_{kl} + \phi)] + (g_{kl} + jb_{kl}) V_k^2 \tag{3.27}$$

Separating the real and imaginary part of Eq. (3.27), we obtain the active and reactive flows:

$$\begin{aligned} P_{kl} &= g_{kl} V_k^2 - V_k V_l g_{kl} \cos(\theta_{kl} + \phi) - V_k V_l b_{kl} \sin(\theta_{kl} + \phi) \\ Q_{kl} &= -b_{kl} V_k^2 + V_k V_l b_{kl} \cos(\theta_{kl} + \phi) - V_k V_l g_{kl} \sin(\theta_{kl} + \phi) \end{aligned} \tag{3.28}$$

Following the same procedure described above, we have the complex power flow from bus l to bus k:

$$\begin{aligned} S_{lk}^* &= P_{lk} - jQ_{lk} = E_l^* I_{lk} \\ S_{lk}^* &= E_l^* \left[y_{kl} E_l - y_{kl} E_k e^{j\phi} \right] \\ S_{lk}^* &= V_l e^{-j\theta_l} \left[y_{kl} V_l e^{j\theta_l} - y_{kl} V_k e^{j\theta_k} e^{j\phi} \right] \\ S_{lk}^* &= y_{kl} V_l^2 - y_{kl} V_k V_l e^{-j(\theta_l - \theta_k - \phi)} \end{aligned} \tag{3.29}$$

If $e^{-j(\theta_{lk}-\phi)} = \cos(\theta_{lk}-\phi) - j\sin(\theta_{lk}-\phi)$ and $y_{kl} = g_{kl} + jb_{kl}$, we have

$$S_{lk}^* = (g_{kl} + jb_{kl}) V_l^2 - (g_{kl} + jb_{kl}) V_k V_l [\cos(\theta_{lk} - \phi) - j sin(\theta_{lk} - \phi)] \tag{3.30}$$

Separating the real and imaginary part of Eq. (3.30), we obtain the active and reactive power flows:

$$\begin{aligned} P_{lk} &= g_{kl} V_l^2 - V_k V_l g_{kl} \cos(\theta_{lk} - \phi) - V_k V_l b_{kl} \sin(\theta_{lk} - \phi) \\ Q_{lk} &= -b_{kl} V_l^2 + V_k V_l b_{kl} \cos(\theta_{lk} - \phi) - V_k V_l g_{kl} \sin(\theta_{lk} - \phi) \end{aligned} \tag{3.31}$$

Generalized π model

Usually, the expressions of power flow between buses or in the branches of an EPS are defined from the generalized π model for branch representation of the bus-branch model, shown in Fig. 3.6 (Monticelli, 1999).

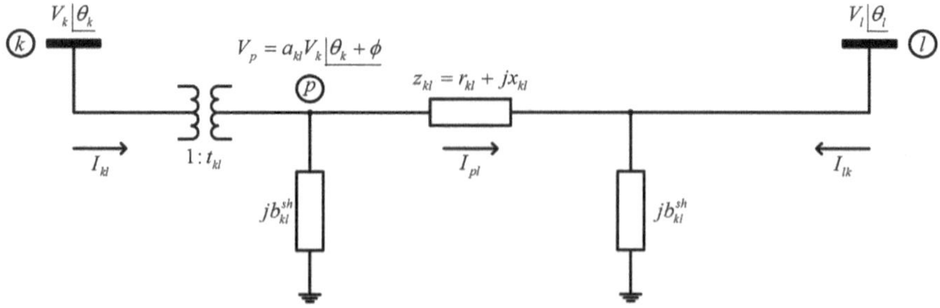

Fig. 3.6 Generalized π model.

Table 3.2 Values of variables a_{kl}, ϕ, and b_{kl}^{sh} for each type of device.

Device	a_{kl}	ϕ	b_{kl}^{sh}
Transmission line	1	0	Any value
Phase-shifting transformer	Any value	Any value	0
In-phase transformer	Any value	0	0
Purely phase-shifting transformer	1	Any value	0

From the application of Kirchhoff's laws on the generalized π model, the following power flow equations are obtained:
(i) From bus k to bus l:

$$P_{kl} = a_{kl}^2 V_k^2 g_{kl} - a_{kl} V_k V_l g_{kl} \cos(\theta_{kl} + \phi) - a_{kl} V_k V_l b_{kl} \sin(\theta_{kl} + \phi) \quad (3.32)$$

$$Q_{kl} = -a_{kl}^2 V_k^2 \left(b_{kl} + b_{kl}^{sh}\right) + a_{kl} V_k V_l b_{kl} \cos(\theta_{kl} + \phi) - a_{kl} V_k V_l g_{kl} \sin(\theta_{kl} + \phi) \quad (3.33)$$

(ii) From bus l to bus k:

$$P_{lk} = V_l^2 g_{kl} - a_{kl} V_k V_l g_{kl} \cos(\theta_{lk} - \phi) - a_{kl} V_k V_l b_{kl} \sin(\theta_{lk} - \phi) \quad (3.34)$$

$$Q_{lk} = -V_l^2 \left(b_{kl} + b_{kl}^{sh}\right) + a_{kl} V_k V_l b_{kl} \cos(\theta_{lk} - \phi) - a_{kl} V_k V_l g_{kl} \sin(\theta_{lk} - \phi) \quad (3.35)$$

Note 3.3: Note that the effect of the transformer is related to the bus k, this occurs because the transformer is connected to this bus. Thus it is extremely important to note that the a_{lk} ratio is not part of the equation, so due care must be taken when implementing the flows from bus l to bus k.

Note 3.4: In the generalized expression of Q_{kl} (Eq. 3.33), the term $-a_{kl}^2$ multiplies b_{kl}^{sh}, which physically does not exist. However, its use is not wrong, since if the device is a transmission line, a_{lk} is unitary ($a_{kl}=1$) and does not erroneously affect the expression of reactive power. If the device involved is a transformer, b_{kl}^{sh} has null value ($b_{kl}^{sh}=0$), causing no problem to the expression. For such equations, the variables a_{kl}, ϕ, and b_{kl}^{sh} take values according to the corresponding equipment, following Table 3.2.

3.3.2 Bus power injection equations

In order to obtain the equation of the power injections in the buses of an EPS, first, it is necessary to obtain the net value of current injection in a generic bus, as shown in Fig. 3.7.

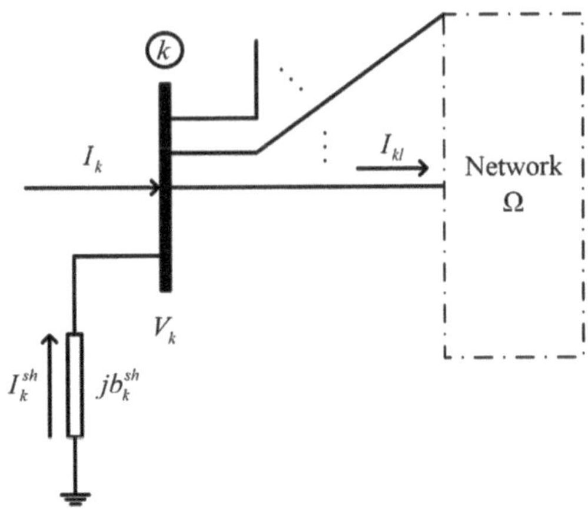

Fig. 3.7 Currents incident to a generic bus *k*.

Thus for the general case, the following equation is valid:

$$I_k + I_k^{sh} = \sum_{l \in \Omega_k} I_{kl} \tag{3.36}$$

for $k = 1, \ldots, n$, where k is a generic bus of EPS, l is a node adjacent to k, Ω_k represents the set of buses adjacent to the bus k, and n is the number of buses of the system.

From the nodal analysis of the generalized π model illustrated in Fig. 3.6, we have the following generalized expression for the current flow from bus k to bus l:

$$I_{kl} = \left(a_{kl}^2 y_{kl} + jb_{kl}^{sh}\right) E_k + \left(-a_{kl} y_{kl} e^{-j\phi}\right) E_l \tag{3.37}$$

Thus I_k can be rewritten as

$$I_k = \left[jb_k^{sh} + \sum_{l \in \Omega_k} \left(a_{kl}^2 y_{kl} + jb_{kl}^{sh}\right) \right] E_k + \sum_{l \in \Omega_k} \left(-a_{kl} e^{-j\phi} y_{kl}\right) E_l \tag{3.38}$$

Eq. (3.38) can be rearranged in matrix form as

$$I = Y.E \tag{3.39}$$

where:

I is the vector of current injections in the buses;
Y is the admittance matrix of the system; and.
E is the vector of the voltages in the buses (nodal voltages).

The off-diagonal elements of the matrix Y are as follows:

$$Y_{kl} = -a_{kl} e^{-j\phi} y_{kl} \tag{3.40}$$

The elements of the main diagonal of Y are

$$Y_{kk} = y_k^{sh} + \sum_{l \in \Omega_k} \left(a_{kl}^2 y_{kl} + y_{kl}^{sh} \right) \tag{3.41}$$

Thus the complete matrix form of current injections in a generic bus k is given by

$$I_k = Y_{kk} E_k + \sum_{l \in \Omega_k} Y_{kl} E_l = \sum_{l \in K} Y_{kl} E_l \tag{3.42}$$

where K is the number of buses adjacent to bus k, including itself.

The matrix Y is commonly decomposed into real and imaginary parts, which are represented, respectively, by G and B, or rather, $Y_{kl} = G_{kl} + jB_{kl}$. Thus the expression of the resulting current injection becomes

$$I_k = \sum_{l \in K} (G_{kl} + jB_{kl}) E_l \tag{3.43}$$

From the expression of the complex power injection in a bus, the following equation follows:

$$S_k^* = P_k - jQ_k = E_k^* I_k$$

$$S_k^* = V_k e^{-j\theta_k} \left\{ \left[jb_k^{sh} + \sum_{l \in \Omega_k} \left(a_{kl}^2 y_{kl} + jb_{kl}^{sh} \right) \right] E_k + \sum_{l \in \Omega_k} \left(-a_{kl} e^{-j\phi} y_{kl} \right) E_l \right\}$$

$$S_k^* = V_k^2 \left[jb_k^{sh} + \sum_{l \in \Omega_k} \left(a_{kl}^2 y_{kl} + jb_{kl}^{sh} \right) \right] + V_k V_l \sum_{l \in \Omega_k} \left(-a_{kl} e^{-j(\theta_k - \theta_l + \phi)} y_{kl} \right)$$

$$S_k^* = V_k^2 jb_k^{sh} + V_k^2 \sum_{l \in \Omega_k} \left[a_{kl}^2 (g_{kl} + jb_{kl}) + jb_{kl}^{sh} \right]$$

$$+ \ldots - V_k V_l \sum_{l \in \Omega_k} \left\{ a_{kl} (g_{kl} + jb_{kl}) \left[\cos(\theta_{kl} + \phi) - j sin(\theta_{kl} + \phi) \right] \right\} \tag{3.44}$$

Isolating the real part from the imaginary part, one has

$$P_k = \sum_{l \in \Omega_k} V_k^2 a_{kl}^2 g_{kl} - \sum_{l \in \Omega_k} a_{kl} V_k V_l g_{kl} \cos(\theta_{kl} + \phi) - \sum_{l \in \Omega_k} a_{kl} V_k V_l b_{kl} \sin(\theta_{kl} + \phi)$$

$$Q_k = -V_k^2 b_k^{sh} - \sum_{l \in \Omega_k} V_k^2 \left(a_{kl}^2 b_{kl} + b_{kl}^{sh} \right) + \sum_{l \in \Omega_k} a_{kl} V_k V_l b_{kl} \cos(\theta_{kl} + \phi) + \ldots$$

$$-\sum_{l \in \Omega_k} a_{kl} V_k V_l g_{kl} \sin(\theta_{kl} + \phi) \tag{3.45}$$

In terms of the admittance matrix Y, the above equations are given by

$$P_k = V_k \sum_{l \in \Omega k} V_l (G_{kl} \cos \theta_{kl} + B_{kl} \sin \theta_{kl}) = \sum_{l \in \Omega k} P_{kl} \tag{3.46}$$

$$Q_k = V_k \sum_{l \in \Omega_k} V_l (G_{kl} \sin \theta_{kl} - B_{kl} \cos \theta_{kl}) = -V_k^2 b_k^{sh} + \sum_{l \in \Omega_k} Q_{kl} \tag{3.47}$$

For more details on the complete equations of the expressions of flows and power injections, see Monticelli (1999).

3.4 Nonlinear power flow

According to what was presented in the previous section, the basic power flow equations were derived through the application of Kirchhoff laws, resulting in

$$\begin{aligned} P_k &= V_k \sum_{l \in \Omega_k} V_l (G_{kl} \cos \theta_{kl} + B_{kl} \sin \theta_{kl}) \\ Q_k &= V_k \sum_{l \in \Omega_k} V_l (G_{kl} \sin \theta_{kl} - B_{kl} \cos \theta_{kl}) \end{aligned} \tag{3.48}$$

Usually, the solution to the power flow problem involves two steps, which can be solved simultaneously or alternately. The first, main subject of this chapter and known as basic solution, is related with the solution through iterative methods of a system of algebraic equations like (3.48). The second one considers the actuation of control systems and the representation of operational limits of the system.

The procedure for obtaining the basic solution to the power flow problem, i.e., with no consideration of the operational constraints nor action of control systems, can be decomposed in two systems of algebraic equations. Considering that NPQ and NPV are, respectively, the number of PQ and PV buses of the network, the subsystems are the following:

1. Subsystem 1 (dimension $2NPQ+NPV$): P_k and Q_k are given in the PQ buses and P_k in the PV buses. The unknowns are V_k and θ_k in the buses PQ and θ_k in the PV buses, so it is a system of $2NPQ+NPV$ nonlinear algebraic equations with the same number of unknowns, described below:

$$\begin{aligned} P_k^{sp} - V_k \sum_{l \in \Omega_k} V_l (G_{kl} \cos \theta_{kl} + B_{kl} \sin \theta_{kl}) = 0 \\ Q_k^{sp} - V_k \sum_{l \in \Omega_k} V_l (G_{kl} \sin \theta_{kl} - B_{kl} \cos \theta_{kl}) = 0 \end{aligned} \tag{3.49}$$

where the superscript "*sp*" denotes specified value. In this case, the unknowns are implicit, which requires an iterative solution process.

2. Subsystem 2 (dimension $NPV+2$): After the solution of Subsystem 1, V_k and θ_k are known for all the buses. Now, one must calculate P_k and Q_k on the reference bus and Q_k on the PV buses. It is a system of $NPV+2$ nonlinear algebraic equations with the same number of unknowns, described by

$$\begin{aligned} P_k &= V_k \sum_{l \in \Omega_k} V_l (G_{kl} \cos\theta_{kl} + B_{kl} \sin\theta_{kl}) \\ Q_k &= V_k \sum_{l \in \Omega_k} V_l (G_{kl} \sin\theta_{kl} - B_{kl} \cos\theta_{kl}) \end{aligned} \tag{3.50}$$

In this case, all the unknowns appear explicitly, which makes the solution process trivial.

The unknowns of Subsystem 1 can be grouped in vector x, given below:

$$x = \begin{bmatrix} \theta \\ V \end{bmatrix} \tag{3.51}$$

where θ is the vector of the voltage angles of the buses PQ and PV and V is the vector of magnitudes of the voltages of the buses PQ. The expressions in (3.49) can be rewritten as

$$\begin{aligned} \Delta P_k &= P_k^{sp} - P_k(V, \theta) = 0 \\ \Delta Q_k &= Q_k^{sp} - Q_k(V, \theta) = 0 \end{aligned} \tag{3.52}$$

or, in vector form:

$$\begin{aligned} \Delta P &= P^{sp} - P(V, \theta) = 0 \\ \Delta Q &= Q^{sp} - Q(V, \theta) = 0 \end{aligned} \tag{3.53}$$

where P is the vector of the active power injections in the buses PQ and PV, and Q is the vector of the reactive power injections in the buses PQ.

Let $g(x)$ be the vector function given by

$$g(x) = \begin{bmatrix} \Delta P \\ \Delta Q \end{bmatrix} \tag{3.54}$$

Through this function, Subsystem 1, given by (3.52), can be placed in the form $g(x) = 0$. This system of nonlinear algebraic equations can be solved by several methods, the most efficient being Newton's and their decoupled variants.

3.4.1 Newton method

Consider an n-dimensional system of type (Monticelli, 1999):

$$g(x) = 0 \tag{3.55}$$

in which $g(x)$ is a vector function of dimension $(n \times 1)$ and x is the vector of unknowns $(n \times 1)$, that is

$$\begin{aligned} g(x) &= [g_1(x), \ldots, g_n(x)]^T \\ x &= [x_1, \ldots, x_n]^T \end{aligned} \tag{3.56}$$

It is desired to determine the value of x for which the function $g(x)$ is equal to zero. Geometrically, as shown in Fig. 3.8 for the one-dimensional case, the solution of (3.55) corresponds to the point where the curve intersects the x-axis. The central idea of the method is to start from an initial approximation to the root (x^0) of the function under analysis to sequentially refine this approximation through an iterative process.

Linearizing the vector function $g(x)$ through its Taylor series expansion around a starting point for $x = x^\nu$ and neglecting the higher order terms, one has

$$g(x^\nu + \Delta x^\nu) \cong g(x^\nu) + J(x^\nu)\Delta x^\nu \tag{3.57}$$

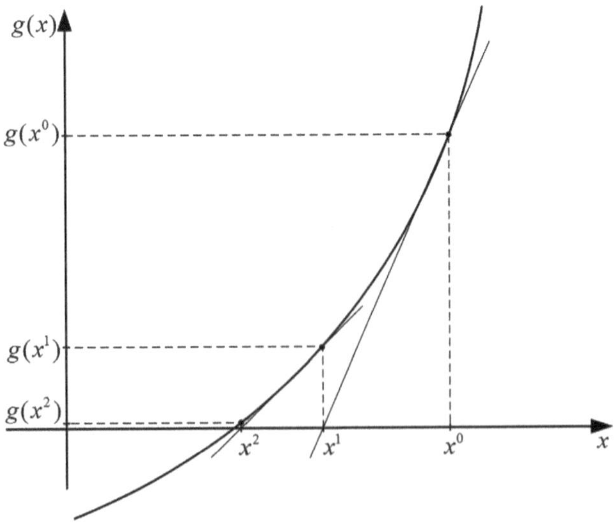

Fig. 3.8 Newton's method for the one-dimensional case.

where the Jacobian matrix J is given by

$$J = \frac{\partial g}{\partial x} = \begin{bmatrix} \frac{\partial g_1}{\partial x_1} & \frac{\partial g_1}{\partial x_2} & \cdots & \frac{\partial g_1}{\partial x_n} \\ \frac{\partial g_2}{\partial x_1} & \frac{\partial g_2}{\partial x_2} & \cdots & \frac{\partial g_2}{\partial x_n} \\ \vdots & \vdots & \ddots & \vdots \\ \frac{\partial g_n}{\partial x_1} & \frac{\partial g_n}{\partial x_2} & \cdots & \frac{\partial g_n}{\partial x_n} \end{bmatrix} \tag{3.58}$$

The correction vector Δx is calculated by

$$g(x^\nu) + J(x^\nu)\Delta x^\nu = 0 \tag{3.59}$$

which is the linearized form of solving the problem $g(x^\nu + \Delta x^\nu) = 0$.

3.4.2 Power flow solution by Newton's method

The Newton method can be applied for the solution of Subsystem 1, as presented in Tinney and Hart (1967). The main aspect of the solution process is to determine the correction vector Δx, which requires the solution of the linear system (3.59) that can be rewritten as

$$g(x^\nu) = -J(x^\nu)\Delta x^\nu \tag{3.60}$$

Subsystem 1 to be solved can be represented by the following system of equations:

$$g(x^\nu) = \begin{bmatrix} \Delta P^\nu \\ \Delta Q^\nu \end{bmatrix}$$

$$\Delta x^\nu = \begin{bmatrix} \Delta \theta^\nu \\ \Delta V^\nu \end{bmatrix}$$

$$J(x^\nu) = \begin{bmatrix} \frac{\partial(\Delta P)}{\partial \theta} & \frac{\partial(\Delta P)}{\partial V} \\ \frac{\partial(\Delta Q)}{\partial \theta} & \frac{\partial(\Delta Q)}{\partial V} \end{bmatrix}^{(\nu)} \tag{3.61}$$

Considering the expressions of the vectors ΔP and ΔQ given in (3.53) and the fact that P^{sp} and Q^{sp} are constants (so their derivatives are null), the Jacobian matrix can be rewritten as:

$$J_{kk} = \frac{\partial g_k}{\partial x_k} \rightarrow J_{11} = \frac{\partial \Delta P_1}{\partial \theta_1}$$

$$\frac{\partial \Delta P_1}{\partial \theta_1} = \frac{\partial (P_1^{sp} - P_1)}{\partial \theta_1} = -\frac{\partial P_1}{\partial \theta_1} \tag{3.62}$$

$$J(x^{\nu}) = -\begin{bmatrix} \frac{\partial P}{\partial \theta} & \frac{\partial P}{\partial V} \\ \frac{\partial Q}{\partial \theta} & \frac{\partial Q}{\partial V} \end{bmatrix}^{(\nu)}$$

Finally, (3.60) can be put in the format:

$$\begin{bmatrix} \Delta P^{\nu} \\ \Delta Q^{\nu} \end{bmatrix} = \begin{bmatrix} \frac{\partial P}{\partial \theta} & \frac{\partial P}{\partial V} \\ \frac{\partial Q}{\partial \theta} & \frac{\partial Q}{\partial V} \end{bmatrix}^{(\nu)} \cdot \begin{bmatrix} \Delta \theta^{\nu} \\ \Delta V^{\nu} \end{bmatrix} \tag{3.63}$$

The components of the Jacobian submatrices are given by

$$\frac{\partial P_k}{\partial \theta_l} = V_k V_l (G_{kl} \sin\theta_{kl} - B_{kl} \cos\theta_{kl})$$
$$\frac{\partial P_k}{\partial \theta_k} = -V_k^2 B_{kk} - V_k \sum_{l \in \Omega_k} V_l (G_{kl} \sin\theta_{kl} - B_{kl} \cos\theta_{kl}) \tag{3.64}$$

$$\frac{\partial P_k}{\partial V_l} = V_k (G_{kl} \cos\theta_{kl} + B_{kl} \sin\theta_{kl})$$
$$\frac{\partial P_k}{\partial V_k} = V_k G_{kk} + \sum_{l \in \Omega_k} V_l (G_{kl} \cos\theta_{kl} + B_{kl} \sin\theta_{kl}) \tag{3.65}$$

$$\frac{\partial Q_k}{\partial \theta_l} = -V_k V_l (G_{kl} \cos\theta_{kl} + B_{kl} \sin\theta_{kl})$$
$$\frac{\partial Q_k}{\partial \theta_k} = -V_k^2 G_{kk} + V_k \sum_{l \in \Omega_k} V_l (G_{kl} \cos\theta_{kl} + B_{kl} \sin\theta_{kl}) \tag{3.66}$$

$$\frac{\partial Q_k}{\partial V_l} = V_k (G_{kl} \sin\theta_{kl} - B_{kl} \cos\theta_{kl})$$
$$\frac{\partial Q_k}{\partial V_k} = -V_k B_{kk} + \sum_{l \in \Omega_k} V_l (G_{kl} \sin\theta_{kl} - B_{kl} \cos\theta_{kl}) \tag{3.67}$$

The diagonal elements can be represented as a function of the active and reactive power injections at bus k, as can be deduced from expressions (3.64) to (3.67):

$$\frac{\partial P_k}{\partial \theta_k} = -Q_k - V_k^2 B_{kk} \tag{3.68}$$

$$\frac{\partial P_k}{\partial V_k} = V_k^{-1}\left(P_k + V_k^2 G_{kk}\right) \tag{3.69}$$

$$\frac{\partial Q_k}{\partial \theta_k} = P_k - V_k^2 G_{kk} \tag{3.70}$$

$$\frac{\partial Q_k}{\partial V_k} = V_k^{-1}\left(Q_k - V_k^2 B_{kk}\right) \tag{3.71}$$

The Newton method applied to the solution of Subsystem 1 is described in the following (Monticelli, 1999):

1. Set $\nu = 0$ and choose the initial values of the angles of the buses PQ and PV ($\theta = \theta^\nu$) and the magnitudes of the buses PQ ($V = V^\nu$);
2. Calculate P_k (V^ν, θ^ν) for the buses PQ and PV, and Q_k (V^ν, θ^ν) for the buses PQ and determine ΔP_k^ν and ΔQ_k^ν;
3. Test for the convergence criterion: if $\max\{\Delta P_k^\nu\} \leq \varepsilon_P$ and $\max\{\Delta Q_k^\nu\} \leq \varepsilon_Q$, then the iterative process has converged to the solution (V^ν, θ^ν). Otherwise, proceed to step 4.
4. Calculate the Jacobian matrix:

$$J(\theta^\nu, V^\nu) = \begin{bmatrix} \dfrac{\partial P}{\partial \theta} & \dfrac{\partial P}{\partial V} \\ \dfrac{\partial Q}{\partial \theta} & \dfrac{\partial Q}{\partial V} \end{bmatrix}_{(\theta^\nu,\, V^\nu)} \tag{3.72}$$

5. Determine the new solution $(V^{\nu+1}, \theta^{\nu+1})$:

$$\theta^{\nu+1} = \theta^\nu + \Delta\theta^\nu \tag{3.73}$$

$$V^{\nu+1} = V^\nu + \Delta V^\nu \tag{3.74}$$

where $\Delta\theta^\nu$ and ΔV^ν are determined by solving the linear system given by

$$\begin{bmatrix} \Delta\theta^\nu \\ \Delta V^\nu \end{bmatrix} = -[J(\theta^\nu, V^\nu)]^{-1} \cdot \begin{bmatrix} \Delta P^\nu \\ \Delta Q^\nu \end{bmatrix} \tag{3.75}$$

6. Set $\nu + 1 \rightarrow \nu$ and return to step 2.

Example 3.1

Consider the two-bus system shown in Fig. 3.9. In this example, the solution of Subsystem 1 is reduced to the calculation of angle θ_2, because the remaining unknowns (θ_1, V_1 and V_2) are specified for this example, since bus 1 is of type $V\theta$ and bus 2 of the PV type. Bus data are shown in Table 3.3 and branch data are shown in Table 3.4 (pu values). Consider the tolerance for convergence $\varepsilon = 0.001$.

Solution: The system model is given by

$$z_{12} = r_{12} + jx_{12} = 0.16 + j0.96$$

$$y_{12} = g_{12} + jb_{12} = \frac{1}{z_{12}} = 0.1689 - j1.0135$$

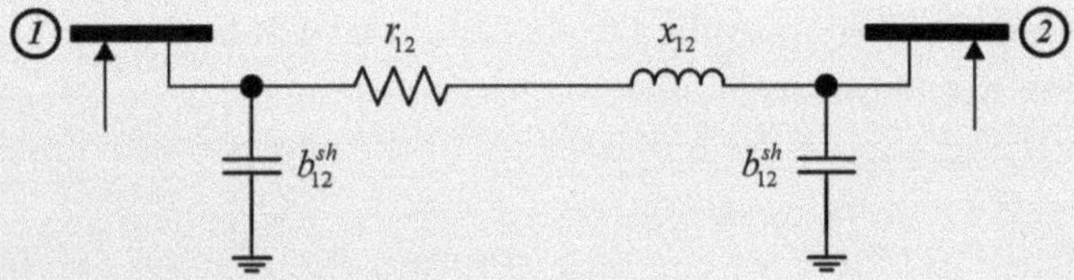

Fig. 3.9 Example system of two buses.

Table 3.3 Bus data (pu values).

Bus	Type	P	Q	Voltage	θ
1	$V\theta$	–	–	1.00	0.00
2	PV	−0.48	–	1.00	–

Table 3.4 Branch data (pu values).

From	To	r	x	b^{sh}
1	2	0.16	0.96	0.02

The nodal admittance matrix is given by

$$Y_{kl} = -y_{kl}$$

$$Y_{kk} = \sum_{l \neq k} \left(y_{kl} + y_{kl}^{sh} \right)$$

$$Y = \begin{bmatrix} 0.1689 - j0.9935 & -0.1689 + j1.0135 \\ -0.1689 + j1.0135 & 0.1689 - j0.9935 \end{bmatrix}$$

Therefore

$$G = real(Y) = \begin{bmatrix} 0.1689 & -0.1689 \\ -0.1689 & 0.1689 \end{bmatrix}$$

$$B = imag(Y) = \begin{bmatrix} -j0.9935 & j1.0135 \\ j1.0135 & -j0.9935 \end{bmatrix}$$

- Subsystem 1 solution:

The initial conditions are given by the flat start, i.e.: $V_1 = V_2 = 1.0$ pu $\theta_1 = \theta_2 = 0.0$ rad. With these data, convergence is tested in relation to the initial conditions (iteration $v = 0$):

$$P_k = V_k \sum_{l=1}^{2} V_l (G_{kl} \cos\theta_{kl} + B_{kl} \sin\theta_{kl})$$

$$P_2^{(0)} = V_2[V_1(G_{21}\cos\theta_{21} + B_{21}\sin\theta_{21}) + V_2(G_{22}\cos\theta_{22} + B_{22}\sin\theta_{22})]$$

$$P_2^{(0)} = V_2[V_1(G_{21}\cos\theta_{21} + B_{21}\sin\theta_{21}) + V_2 G_{22}]$$

$$P_2^{(0)} = -0.1689\cos 0 + 1.0135\sin 0 + 0.1689 = 0$$

$$\Delta P_2^{(0)} = P_2^{sp} - P_2^{(0)} = -0.48 - 0.00 = -0.48$$

$$|\Delta P_2| > \varepsilon$$

Since there was no convergence, start the iterative process by the Newton method.

First iteration: $v = v + 1 = 1$.

$$\Delta P_2^{(0)} = -J\Delta\theta_2^{(1)}$$

$$J^{(1)} = \frac{\partial P_2}{\partial \theta_2}$$

$$J^{(1)} = -V_2^2 B_{22} - V_2[V_1(G_{21}\sin\theta_{21} - B_{21}\cos\theta_{21}) + V_2(G_{22}\sin\theta_{22} + B_{22}\cos\theta_{22})]$$

$$J^{(1)} = -V_2^2 B_{22} - V_2[V_1(G_{21}\sin\theta_{21} - B_{21}\cos\theta_{21}) + V_2 B_{22}]$$

$$J^{(1)} = 0.9935 + 0.1689\sin 0 + 1.0135\cos 0 - 0.9935 = 1.0135$$

$$\Delta\theta_2^{(1)} = J^{(1)^{-1}} \Delta P_2^{(0)} = -\frac{0.48}{1.0135} = -0.4736$$

$$\theta_2^{(1)} = \theta_2^{(0)} + \Delta\theta_2^{(1)} = 0 - 0.4736 = -0.4736$$

$$P_2^{(1)} = V_2[V_1(G_{21}\cos\theta_{21} + B_{21}\sin\theta_{21}) + V_2G_{22}]$$

$$P_2^{(1)} = -0.1689\cos -0.4736 + 1.0135\sin -0.4736 + 0.1689 = -0.4437$$

$$\Delta P_2^{(1)} = P_2^{sp} - P_2^{(1)} = -0.48 + 0.4437 = -0.0363$$

$$|\Delta P_2| > \varepsilon$$

Since there was no convergence, continue the iterative process.
Second iteration: $v = v + 1 = 2$.

$$J^{(2)} = 0.9935 + 0.1689\sin -0.4436 + 1.0135\cos -0.4436 - 0.9935$$
$$= 0.8249$$

$$\Delta\theta_2^{(2)} = J^{-1}\Delta P_2^{(1)} = -\frac{-0.0363}{0.8249} = -0.0440$$

$$\theta_2^{(2)} = \theta_2^{(1)} + \Delta\theta_2^{(2)} = -0.4736 - 0.0440 = -0.5176$$

$$P_2^{(2)} = -0.1689\cos -0.5176 + 1.0135\sin -0.5176 + 0.1689 = -0.4794$$

$$\Delta P_2^{(2)} = P_2^{sp} - P_2^{(2)} = -0.48 + 0.4794 = -0.0006$$

$$|\Delta P_2| < \varepsilon$$

So, the system converged to the solution $\theta_2 = -0.5176$ rad.

- Subsystem 2 solution:

$$P_k = V_k \sum_{l\in\Omega_k} V_l(G_{kl}\cos\theta_{kl} + B_{kl}\sin\theta_{kl})$$

$$P_1 = V_1[V_1(G_{11}\cos\theta_{11} + B_{11}\sin\theta_{11}) + V_2(G_{12}\cos\theta_{12} + B_{12}\sin\theta_{12})]$$

$$P_1 = V_1[V_1G_{11} + V_2(G_{12}\cos\theta_{12} + B_{12}\sin\theta_{12})]$$

$$P_1 = 0,1689 - 0,1689\cos(0,5176) + 1,0135\sin(0,5176)$$

$$P_1 = 0,5237\,pu$$

$$Q_k = V_k \sum_{l\in\Omega_k} V_l(G_{kl}\sin\theta_{kl} - B_{kl}\cos\theta_{kl})$$

$$Q_1 = V_1[V_1(G_{11}\sin\theta_{11} - B_{11}\cos\theta_{11}) + V_2(G_{12}\sin\theta_{12} - B_{12}\cos\theta_{12})]$$

$$Q_1 = V_1[-V_1B_{11} + V_2(G_{12}\sin\theta_{12} - B_{12}\cos\theta_{12})]$$

$$Q_1 = 0,9935 - 0,1689\sin(0,5176) - 1,0135\cos(0,5176)$$

$$Q_1 = 0,0292\,pu$$

$$Q_2 = V_2[V_1(G_{21}\sin\theta_{21} - B_{21}\cos\theta_{21}) + V_2(G_{22}\sin\theta_{22} - B_{22}\cos\theta_{22})]$$

$$Q_2 = V_2[V_1(G_{21}\sin\theta_{21} - B_{21}\cos\theta_{21}) - V_2B_{22}]$$

$$Q_2 = -0,1689\sin(-0,5176) - 1,0135\cos(-0,5176) + 0,9935$$

$$Q_2 = 0,1964\,pu$$

3.5 Power flow in distribution systems

The traditional methods for calculating power flow for transmission systems, such as Newton's method presented in the previous section, are generally not suitable for calculating power flow in distribution systems. This assertion is justified by the fact that these methods require the factorization of matrices that, due to the characteristics of the distribution systems (low X/R (reactance/resistance) ratio of the lines, the connection of low impedance segments with high impedance segments and the large number of buses and lines), are ill-conditioned.

Considering the above statement, specific methods for calculating power flow for distribution systems were developed. The most used methods are forward/backward sweep, since they are robust and simple to implement. Those methods do not require the factorization of matrices and consider the characteristic of the distribution systems to operate in a radial structure, thus increasing the computational efficiency to determine the state of the network (Santos, 2009).

Forward/Backward sweeping methods are named so because they have two distinct steps in the process of obtaining the state of the grid. The process starts with a backward sweep, in which the currents (or powers) of all the segments are calculated, starting from the end of the feeders up to the substation. In this step, it is assumed that the voltages in the buses are their nominal voltages.

Next, the forward sweep is carried out, where all the nodal voltages are calculated using the currents (or powers) obtained in the previous step, the value of the voltage of the substation, and the data of impedance of the lines.

The backward and forward sweeps are performed iteratively, considering the previously obtained values, until a convergence criterion is satisfied.

3.5.1 Power summation method

Power Summation (Baran and Wu, 1989; Cespedes, 1990) is the most efficient backward/forward sweep method in the literature on power distribution systems and the most efficient method of calculating power flow in radial or weakly meshed systems. To illustrate how this method works, consider, for example, the feeder illustrated in Fig. 3.10.

The execution of this method starts with the backward step, of inverse sweep. The power flow that goes to bus l through the branch kl is calculated as follows:

$$S'_{kl} = S_l + S_r \tag{3.76}$$

Separating the terms regarding the active and reactive power in (3.76), we have

$$\begin{aligned} P'_{kl} &= P_l + P_r \\ Q'_{kl} &= Q_l + Q_r \end{aligned} \tag{3.77}$$

Eq. (3.76) can also be written in terms of the power flow leaving the bus k through branch kl as described below:

$$S_{kl} = S'_{kl} + z_{kl}|I_{kl}|^2 \tag{3.78}$$

where $I_{kl} = \left(\frac{S'_{kl}}{V_l}\right)$ and $|I_{kl}|^2 = \frac{(P'_{kl})^2 + (Q'_{kl})^2}{V_l^2}$. Substituting this last expression in (3.78) yields the complex power S_{kl}, given by

$$S_{kl} = P'_{kl} + r_{kl}\left[\frac{(P'_{kl})^2 + (Q'_{kl})^2}{V_l^2}\right] + j\left\{Q'_{kl} + x_{kl}\left[\frac{(P'_{kl})^2 + (Q'_{kl})^2}{V_l^2}\right]\right\} \tag{3.79}$$

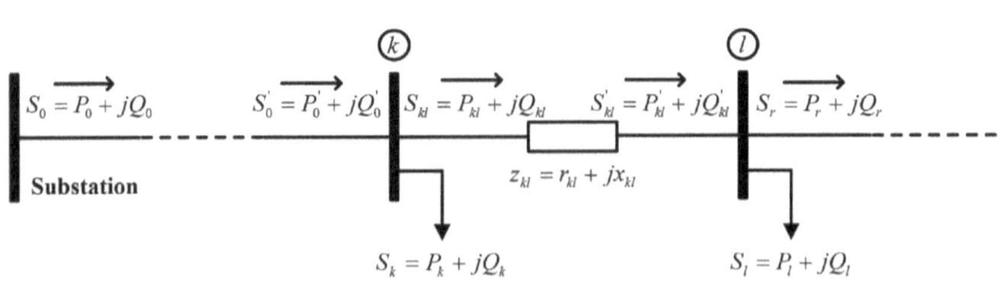

Fig. 3.10 Branches from a feeder of a distribution system.

Separating again the real and imaginary parts, we can determine the active and reactive powers that flow through the branch *kl*:

$$P_{kl} = P'_{kl} + r_{kl}\left[\frac{\left(P'_{kl}\right)^2 + \left(Q'_{kl}\right)^2}{V_l^2}\right]$$
$$Q_{kl} = Q'_{kl} + x_{kl}\left[\frac{\left(P'_{kl}\right)^2 + \left(Q'_{kl}\right)^2}{V_l^2}\right] \tag{3.80}$$

Eqs. (3.76) and (3.80) are used in the backward sweep which starts at the feeders' end nodes and finishes at the substation bus. After this step has been completed, the complex powers (active and reactive) flowing through all the branches are known. Then, proceed to the forward sweep, determining the value of the voltage of bus l, by means of the expressions below:

$$V_l = V_k - I_{kl}z_{kl} = V_k - \frac{S_{kl}^*}{V_k^*}(r_{kl} + jx_{kl})$$
$$V_l = V_k - \frac{P_{kl} - jQ_{kl}}{V_k}(r_{kl} + jx_{kl}) \tag{3.81}$$
$$V_l = \left(V_k - \frac{r_{kl}P_{kl} + x_{kl}Q_{kl}}{V_k}\right) + j\left(\frac{r_{kl}Q_{kl} - x_{kl}P_{kl}}{V_k}\right)$$

Since the voltage V_l is a complex number, it can be expressed both in the rectangular form, as it was presented in Eq. (3.81), and in the polar form, that carries the information of magnitude and phase angle. The following equation shows the expressions of magnitude and phase of V_l:

$$|V_l| = \sqrt{\left(V_k - \frac{r_{kl}P_{kl} + x_{kl}Q_{kl}}{V_k}\right)^2 + \left(\frac{r_{kl}Q_{kl} - x_{kl}P_{kl}}{V_k}\right)^2}$$
$$|V_l| = \sqrt{V_k^2 - 2(r_{kl}P_{kl} + x_{kl}Q_{kl}) + \frac{(r_{kl}P_{kl})^2 + (x_{kl}Q_{kl})^2 + (r_{kl}Q_{kl})^2 + (x_{kl}P_{kl})^2}{V_k^2}} \tag{3.82}$$

$$\theta_l = -\tan^{-1}\left(\frac{r_{kl}Q_{kl} - x_{kl}P_{kl}}{V_k^2 - r_{kl}P_{kl} - x_{kl}Q_{kl}}\right) \tag{3.83}$$

where V_k is the voltage magnitude at bus k. Because it is a forward/backward sweeping method, the process is performed iteratively and is performed in the following steps:

1. Initially, the substation voltage is set to all buses in the system.
2. Through Eqs. (3.76) and (3.80) of the backward sweep, the power flows of all the branches of the network are calculated, starting from the end buses of the network and ending at the substation bus.
3. With the Eqs. (3.82) and (3.83) of the forward sweep, the voltages of all the buses of the system are updated using the data obtained in the previous step. The process starts at the substation bus and is run until all bus voltages are determined.
4. Calculate, for all buses, the difference between the voltage magnitude of the current iteration and the previous one. If the largest difference is greater than a specified tolerance, we return to step 2. Otherwise, the process terminates, and all nodal voltages and power flows of the branches are determined.

3.5.2 Current summation method

The application of the current summation method is more appropriate for radial distribution systems; however, the method can also be applied to weakly meshed systems (Mansour et al., 2009). To explain the operation of the current summation method, consider, for example, the section of a radial feeder with n buses, as shown in Fig. 3.11.

Initially, it is assumed that all the nodal voltages of the network are equal (in magnitude and phase) to the voltage of the substation, which has zero

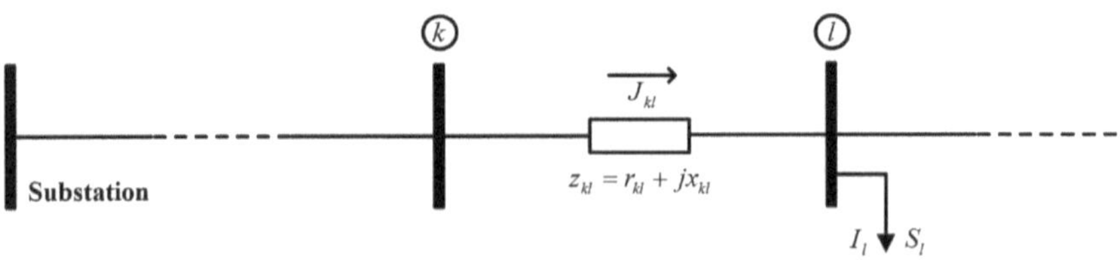

Fig. 3.11 Branches of a radial feeder.

phase angle. The iterative process is then performed according to the following steps:

1. The load currents of each ith bus of the system are calculated by the following equation:

$$I_i^{(\nu)} = \left(\frac{S_i}{V_i^{(\nu-1)}}\right)^* - Y_i^{sh} V_i^{(\nu-1)}, i = 1, \ldots, n, \tag{3.84}$$

where $I_i^{(\nu)}$ is the complex current demanded by the bus i in the iteration ν; $V_i^{(\nu-1)}$ is the complex voltage in bus i in iteration $\nu - 1$; Y_i^{sh} is the sum of all shunt elements connected to bus i; and S_i is the complex power of bus i.

2. Backward sweep: Starting from the end buses of the network and ending at the substation bus, the current $J_{kl}^{(\nu)}$ that circulates in each branch kl connecting a bus l with its predecessor k is calculated by the following equation:

$$J_{kl}^{(\nu)} = I_l^{(\nu)} + \sum_{m \in M} J_{lm}^{(\nu)} \tag{3.85}$$

where M represents the set of all the lines connected to bus l.

3. Forward sweep: Starting from the substation bus (zero depth) and moving to the buses of greater depth (end nodes of the network), the voltage of each bus l of the system, connected to a bus k that is its predecessor (one unit less deep), is calculated by the following equation:

$$V_l^{(\nu)} = V_k^{(\nu)} - z_{kl} J_{kl}^{(\nu)} \tag{3.86}$$

4. In this step, it is verified if the convergence was reached. The largest difference in value of some variable (nodal voltage, for example) between the current and previous iteration is determined. If this difference is greater than the tolerance specified, we return to step 2. Otherwise, the process has converged and all the voltages in the buses and currents in the branches are determined.

It is important to emphasize that in the equation of the two methods presented were not considered current injections entering the buses that were not buses representing substations, that is, it was not considered the existence of generating units connected along the feeders. In this situation, due to the radial configuration of the distribution systems, the direction of the power flow in the network is unique, from the substation to the load buses.

3.6 Exercises

1. Consider the IEEE 14-bus system shown in Fig. 3.12. The bus data are presented in Table 3.5 and the branch data in Table 3.6 (values of P in MW, Q, and b^{sh} in MVAr, magnitude of nodal voltage in Volts, and phase angle of nodal voltage θ in radians).

   With these data, following are requested:

   (a) Calculate the power flow of the IEEE 14-bus system.

   (b) Increase the powers P and Q of bus 13 by multiplying them by a factor of 1.5 and then multiplying them by 2. Comment about the main changes observed.

   (c) Decrease the powers P and Q of bus 13 by dividing them by a factor of 1.5 and then dividing by 2.0. Comment about the main changes observed.

   (d) Suppose lines 2–3 have been removed, what happens to the other lines connected to buses 2 and 3?

2. Using the forward/backward current summation method, determine the nodal complex voltages of the distribution feeder buses shown in Fig. 3.13 (values in pu and degrees). To do this, use the constant power

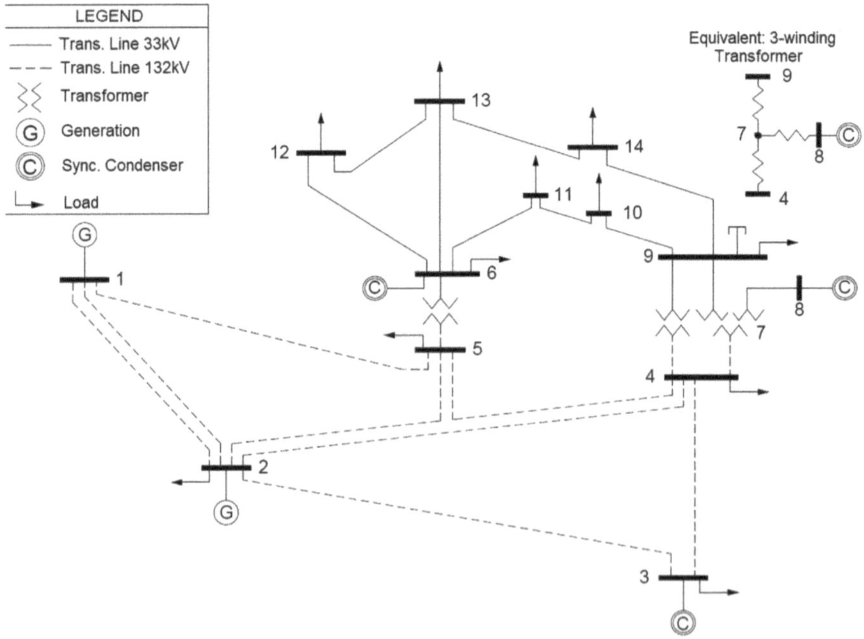

Fig. 3.12 Topology of the IEEE 14-bus system.

Table 3.5 Bus data.

Bus	Type	P (MW)	Q (MVAr)	b^{sh} (MVAr)	Voltage (V)	θ (rad)
1	$V\theta$				1060	0.00
2	PV	18.3			1045	
3	PV	−94.2			1010	
4	PQ	−47.8	3.9			
5	PQ	−7.6	−1.6			
6	PV	−11.2			1070	
7	PQ	0.0	0.0			
8	PV	0.0			1090	
9	PQ	−29.5	−16.6	19		
10	PQ	−9.0	−5.8			
11	PQ	−3.5	−1.8			
12	PQ	−6.1	−1.6			
13	PQ	−13.5	−5.8			
14	PQ	−14.9	−5.0			

Table 3.6 Branches data.

From	To	R (%)	X (%)	b^{sh} (MVAr)	Tap
1	2	1.938	5.917	5.28	
1	5	5.403	22.304	4.92	
2	3	4.699	19.797	4.38	
2	4	5.811	17,632	3.40	
2	5	5.695	17.388	3.46	
3	4	6.701	17.103	1.28	
4	5	1.335	4.211		
4	7		20.912		0.978
4	9		55.618		0.969
5	6		25.202		0.932
6	11	9.498	19.890		
6	12	12.291	25.581		
6	13	6.615	13.027		
7	8		17.615		
7	9		11.001		
9	10	3.181	8.450		
9	14	12.711	27.038		
10	11	8.205	19.207		
12	13	22.092	19.988		
13	14	17.093	34.802		

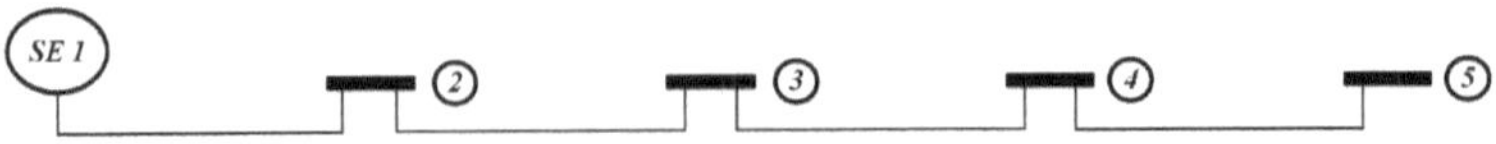

Fig. 3.13 Feeder with five load buses.

Table 3.7 Parameters and loading.

From	To	R (Ω)	X (Ω)	Load on "To" bus	
				P (kW)	Q (kVAr)
1	2	0.3220	0.2700	1000	600
2	3	0.4930	0.2511	900	400
3	4	0.3660	0.1864	1200	800
4	5	0.3811	0.1941	800	600

load model and assume the following base values: $S_{base} = 1000$ kVA and $V_{base} = 12.66$ kV. The parameters and demand of the load buses of this feeder are shown in Table 3.7. Consider as convergence criterion the tolerance of 0.00001.

References

Baran, M.E., Wu, F.F., 1989. Network reconfiguration in distribution systems for loss reduction and load balancing. IEEE Trans. Power Deliv. 4 (2), 1401–1407.

Cespedes, R.G., 1990. New method for the analysis of distribution networks. IEEE Trans. Power Deliv. 5 (1), 391–396.

Mansour, M.R., Santos, A.C., London Jr., J.B.A., Delbem, A.C.B., Bretas, N.G., 2009. Energy Restoration in Distribution Systems Using Multi-Objective Evolutionary Algorithm and an Efficient Data Structure. IEEE Bucharest Powertech, Bucharest, Romania, pp. 1–7.

Monticelli, A., 1999. State Estimation in Electric Power Systems: A Generalized Approach. Kluwer Academic Publishers, Boston, USA.

Santos, A., 2009. Evolutionary Algorithm Computationally Efficient for Distribution System Reconfiguration (Ph.D. Dissertation (Electrical Engineering)). São Carlos School of Engineering (EESC/USP), São Carlos, Brazil.

Sttot, B., 1974. Review of load-flow calculation methods. Proc. IEEE 62 (7), 916–929.

Tinney, W., Hart, C., 1967. Power flow solution by Newton's method. IEEE Trans. Power App. Syst. PAS-86 (11), 1449–1460.

CHAPTER 4

Classical static state estimation in electric power systems

This chapter deals with the third stage of the traditional PSSE process, that is, of the static state estimation proper considering the single-phase modeling of the electric network. As mentioned in Chapter 1, because of the simplicity of its formulation, as well as the ease of its implementation in computer, the WLS estimator is the most used in real-life applications and will be the focus of this chapter. The coupled, decoupled, and linear versions of the WLS estimator will be presented in detail, and some alternative formulations developed to avoid numerical problems. Also, it will be introduced the hybrid WLS estimator, which is the WLS estimator that allows the simultaneous processing of SCADA measurements and SPMs, as well as estimators statistically more robust than the WLS estimator. A section of this chapter will discuss the particularities of the state estimation process in distribution systems (DSs), highlighting some state estimators developed especially for DSs. In the last section are some problems to assist in understanding the WLS estimator.

4.1 Introduction

As mentioned previously, although the generation, transmission, and distribution of electric energy are usually performed through three-phase systems, most of the state estimators proposed in the literature make use of the per-phase or single-phase model (positive sequence model), which is simpler than the three-phase model. It is important to remember that in the single-phase model, it is assumed that the loads are balanced and the networks are symmetrical, that is, the balanced system hypothesis is assumed, where the electrical voltages in the phases have the same magnitude, but with an angular phase shift of 120°. This model is suitable for most ultra high-voltage and extra high-transmission systems, the focus of this book. However, in many situations the imbalances in the network are very accentuated, as is usually the case in DSs, making the use of the per-phase

Cyber-Physical Power Systems State Estimation
https://doi.org/10.1016/B978-0-323-90033-1.00007-X

model impossible (Hansen and Debs, 1995; Zhong and Abur, 2002; Meliopoulos et al., 2005; Almeida et al., 2006).

Considering that the focus of this book is the state estimation process in transmission systems, this chapter will consider the single-phase static state estimator.

In order to introduce this chapter, it is important to remember that the static state estimation, in EPS, consists of obtaining, in real-time, the state variables of an EPS (usually the complex nodal voltages) by processing the following information: network parameters, topology and configuration of the meters in the bus-branch model (obtained by the network configurator), and "snapshots" of real-time analog measurements.

It is important to remember that the formulation of the static state estimation process, which will be presented in this chapter, is based on the modeling of the electric network components and the power flow equations presented in the previous chapter.

4.2 The measurement model

Once the topology and the parameters of an EPS are known, the determination of its state variables is carried out from measured values, which are subject to errors inherent in the measurement process (inaccuracy of the meters, communication channel errors, effects of analog to digital conversion, etc.). Thus the measurements available in an EPS can be related to the measurement errors, through the following model:

$$z = z^{true} + e, \tag{4.1}$$

where z is the vector of the analog measurements of dimension $(m \times 1)$; z^{true} is the vector with the "true" measurement values of dimension $(m \times 1)$; e is the random vector of dimension $(m \times 1)$ representing the measurement errors.

Since the true values of the state variables of an EPS are not known, therefore the true values of the measurements are also unknown. To estimate the state variables, it is necessary to make certain assumptions about the measurement model presented in Eq. (4.1). In the conventional PSSE process, it is assumed that the vector of random errors of the analog measurements has Gaussian distribution with zero mean ($E(e) = 0$) and covariance matrix R ($E(ee^T) = R$), and that errors are uncorrelated. Thus the matrix R is assumed to be diagonal, whose elements are the variances of measurement errors, which are usually expressed as a function of the accuracy and scale of the measuring instruments. From these assumptions, the measurement model can be rewritten as presented in Eq. (4.2):

$$z = h(x) + e, \tag{4.2}$$

where x is the vector of state variables with dimension (N x 1), usually constituted by ($n - 1$) bus voltage phase angles and bus voltage magnitudes, for an EPS with n buses, since one of EPS buses is chosen as angular reference ($N = 2n - 1$); $h(.)$ is the vector of linear and/or nonlinear functions (depending on the type of measurement), relating the measured quantities and the state variables, with dimension ($m \times 1$).

As previously mentioned, usually the analog measurements used in the PSSE process are measurements of active and reactive power (flow in power lines and transformers and bus injection) and magnitude of voltage in buses, obtained by the SCADA system. However, after the development of the phasor measurement units (PMUs), several researches are being carried out to include the synchronous phasor measurements (SPMs) of bus voltage phasors and branch current phasors in the process of PSSE. It should be noted that the time base used in PMUs for SPM synchronization is the global positioning system (GPS) clock signal.

Usually, the information provided by zero-injection buses is also used as measurements in the PSSE process. Such information, defined as virtual measurements (Abur and Expósito, 2004), is accurate, despite the absence of physical measurement. Thus virtual measurements can be treated as low-variance power injection measurements, or as equality constraints to be included in the formulation of the PSSE problem.

4.3 WLS state estimator

In the following sections, the coupled, decoupled, and linearized versions of the WLS estimator will be presented.

4.3.1 Coupled state estimator

Using the classical WLS method, the best estimate of the vector of state variables x, denoted by $\hat{x}$, can be obtained by calculating the value of x that minimizes the index $J(x)$ (Schweppe and Handschin, 1974), given by

$$J(x) = e^T R^{-1} e, \tag{4.3}$$

or

$$J(x) = [z - h(x)]^T R^{-1} [z - h(x)], \tag{4.4}$$

where R^{-1} is the inverse of the covariance matrix of the measurements, used as weight matrix. The covariance matrix R, of dimension $m \times m$, is given by

$$R = \begin{bmatrix} \sigma_1^2 & \cdots & 0 \\ \vdots & \ddots & \vdots \\ 0 & \cdots & \sigma_m^2 \end{bmatrix}, \tag{4.5}$$

where σ_i is the standard deviation of measurement i.

The index $J(x)$ becomes minimal when

$$\frac{\partial J(x)}{\partial x} = 0, \tag{4.6}$$

or equivalently

$$H^T(\hat{x}) R^{-1} [z - h(\hat{x})] = 0, \tag{4.7}$$

where $H(\hat{x})$ is the matrix of the first derivatives of the nonlinear functions of the vector $h(x)$, known as the Jacobian matrix, calculated at the point of estimated state variables $\hat{x}$ and represented by

$$H(\hat{x}) = \left. \frac{\partial h(x)}{\partial x} \right|_{x=\hat{x}}. \tag{4.8}$$

Because the index $J(x)$ is a quadratic nonlinear function, an iterative method is applied to obtain $\hat{x}$ by solving a linear equation at each iteration κ in order to calculate the current estimate of the state variables vector through successive corrections (Schweppe and Handschin, 1974), given by

$$x^{\kappa+1} = x^{\kappa} + \Delta x^{\kappa}. \tag{4.9}$$

However, to determine the correction Δx^{κ}, a linearization of the equations of $h(x)$ is done around the point x^{κ}, represented by the equation:

$$h\left(x^{\kappa+1}\right) \cong h(x^{\kappa}) + H(x^{\kappa}) \Delta x^{\kappa}. \tag{4.10}$$

Rewriting Eq. (4.2), with respect to the approximations performed in $h(x)$, we obtain the linearized measurement model:

$$z = h(x^{\kappa}) + H(x^{\kappa}) \Delta x^{\kappa} + e, \tag{4.11}$$

or:

$$\Delta z(x^{\kappa}) = z - h(x^{\kappa}) = H(x^{\kappa}) \Delta x^{\kappa} + e, \tag{4.12}$$

From the linear measurement model, the objective function $J(\Delta x)$ becomes

$$J(\Delta x) = [\Delta z(x^{\kappa}) - H(x^{\kappa})\Delta x^{\kappa}]^T R^{-1}[\Delta z(x^{\kappa}) - H(x^{\kappa})\Delta x^{\kappa}], \tag{4.13}$$

whose minimum is calculated from

$$\frac{\partial J(\Delta x)}{\partial \Delta x} = H(x^{\kappa})^T R^{-1}[\Delta z(x^{\kappa}) - H(x^{\kappa})\Delta x^{\kappa}] = 0. \tag{4.14}$$

Therefore the solution can be obtained by the following equation:

$$\Delta x^{\kappa} = \left[H(x^{\kappa})^T R^{-1} H(x^{\kappa})\right]^{-1} H(x^{\kappa})^T R^{-1} \Delta z(x^{\kappa}), \tag{4.15}$$

which is called *normal equation*, where

$$H(x^{\kappa})^T R^{-1} H(x^{\kappa}) = G(x^{\kappa}) \tag{4.16}$$

represents the gain matrix (G) calculated in x^{κ}.

The iterative process starts from an initial value x^0 and, at each iteration κ, the corrections in the state variables Δx are obtained through Eq. (4.15). The update of the vector of state variables is obtained through Eq. (4.9) until a stop criterion is satisfied, such as

$$\max|\Delta x^{\kappa}| \leq \varepsilon, \tag{4.17}$$

in which ε denotes a previously established tolerance (corresponds to the degree of acceptance of a solution). Thus this criterion will indicate that the iterative process will be closed when the magnitude of the adjustments in the state variables is negligible.

The coupled WLS state estimator algorithm can be summarized by the steps given below:

- **Step 1:** Make the iteration index $\kappa = 0$ and choose an initial solution $x^{\kappa} = x^0$ (typically as a flat start).
- **Step 2:** Calculate the matrices $H(x^{\kappa})$ and $G(x^{\kappa})$ at the point $x = x^{\kappa}$.
- **Step 3:** Get the correction in the state variables through the normal equation and update the variables:

$$\Delta x^{\kappa} = \left[H(x^{\kappa})^T R^{-1} H(x^{\kappa})\right]^{-1} H(x^{\kappa})^T R^{-1} \Delta z(x^{\kappa}), \tag{4.18}$$

$$x^{\kappa+1} = x^{\kappa} + \Delta x^{\kappa}. \tag{4.19}$$

- **Step 4:** Test the stopping criterion: if $\max \Delta x^{\kappa} \leq \varepsilon$, the process has converged, stop. Otherwise, update $x^{\kappa+1} = x^{\kappa} + \Delta x^{\kappa}$; $\kappa = \kappa + 1$ and return to Step 2.

In addition to the above-mentioned stopping criterion, it is also common to consider the maximum number of iterations (κ_{max}) as the stopping criterion. That is, if after κ_{max} iterations there is no convergence ($\Delta x^{\kappa} > \varepsilon$), the process is interrupted. Fig. 4.1 shows the flowchart of the coupled WLS state estimator algorithm.

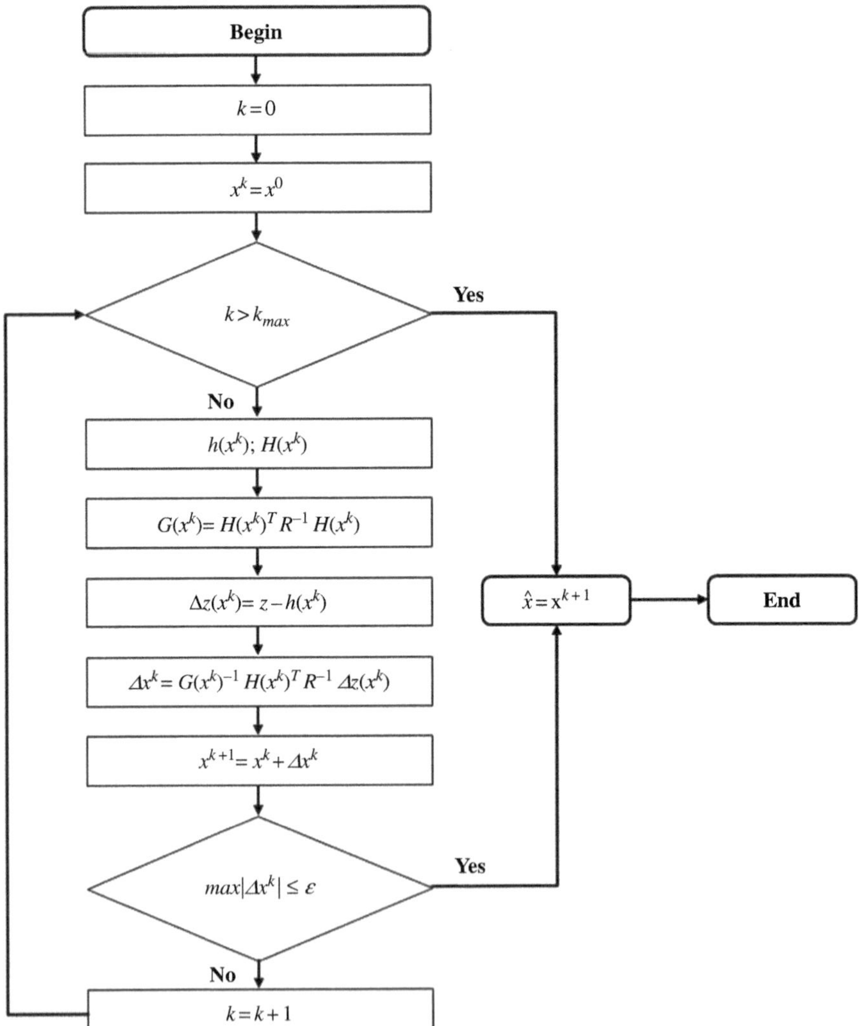

Fig. 4.1 Flowchart of the coupled WLS state estimator algorithm.

4.3.2 Jacobian matrix

The Jacobian matrix (H) relates the measurements available for estimation (matrix rows) with the state variables to be estimated (matrix columns). The Jacobian matrix structure will depend on the order in which the measurements and state variables are arranged in the matrix. Thus it is at the discretion of each one of the positioning of each partial derivative of measurement in relation to the measured state. A possible structure of the Jacobian matrix is presented in the following.

$$H = \begin{bmatrix} \frac{\partial P_{kl}}{\partial \theta_k} & \frac{\partial P_{kl}}{\partial \theta_l} & \frac{\partial P_{kl}}{\partial v_k} & \frac{\partial P_{kl}}{\partial v_l} \\ \frac{\partial Q_{kl}}{\partial \theta_k} & \frac{\partial Q_{kl}}{\partial \theta_l} & \frac{\partial Q_{kl}}{\partial v_k} & \frac{\partial Q_{kl}}{\partial v_l} \\ \frac{\partial P_k}{\partial \theta_k} & \frac{\partial P_k}{\partial \theta_l} & \frac{\partial P_k}{\partial v_k} & \frac{\partial P_k}{\partial v_l} \\ \frac{\partial Q_k}{\partial \theta_k} & \frac{\partial Q_k}{\partial \theta_l} & \frac{\partial Q_k}{\partial v_k} & \frac{\partial Q_k}{\partial v_l} \\ \frac{\partial V_k}{\partial \theta_k} & \frac{\partial V_k}{\partial \theta_l} & \frac{\partial V_k}{\partial v_k} & \frac{\partial V_k}{\partial v_l} \end{bmatrix} \tag{4.20}$$

Based on the power flow, power injection, and voltage magnitude equations developed in Chapter 3 assuming the generalized π model, and the state variables θ_k, θ_l, v_k, and v_l, the equations of the partial derivatives that make up the Jacobian matrix are given as:

- Active power flows from bus k to bus l:

$$\frac{\partial P_{kl}}{\partial \theta_k} = a_{kl} V_k V_l g_{kl} \sin(\theta_{kl} + \varphi_{kl}) - a_{kl} V_k V_l b_{kl} \cos(\theta_{kl} + \varphi_{kl})$$

$$\frac{\partial P_{kl}}{\partial \theta_l} = -a_{kl} V_k V_l g_{kl} \sin(\theta_{kl} + \varphi_{kl}) + a_{kl} V_k V_l b_{kl} \cos(\theta_{kl} + \varphi_{kl})$$

$$\frac{\partial P_{kl}}{\partial V_k} = 2a_{kl}^2 V_k g_{kl} - a_{kl} V_l g_{kl} \cos(\theta_{kl} + \varphi_{kl}) - a_{kl} V_l b_{kl} \sin(\theta_{kl} + \varphi_{kl})$$

$$\frac{\partial P_{kl}}{\partial V_l} = -a_{kl} V_k g_{kl} \cos(\theta_{kl} + \varphi_{kl}) - a_{kl} V_k b_{kl} \sin(\theta_{kl} + \varphi_{kl}). \tag{4.21}$$

- Active power flows from bus l to bus k:

$$\frac{\partial P_{lk}}{\partial \theta_k} = -a_{kl} V_k V_l g_{kl} \sin(\theta_{lk} - \varphi_{kl}) + a_{kl} V_k V_l b_{kl} \cos(\theta_{lk} - \varphi_{kl})$$

$$\frac{\partial P_{lk}}{\partial \theta_l} = a_{kl} V_k V_l g_{kl} \sin(\theta_{lk} - \varphi_{kl}) - a_{kl} V_k V_l b_{kl} \cos(\theta_{lk} - \varphi_{kl})$$

$$\frac{\partial P_{lk}}{\partial V_k} = -a_{kl} V_l g_{kl} \cos(\theta_{lk} - \varphi_{kl}) - a_{kl} V_l b_{kl} \sin(\theta_{lk} - \varphi_{kl})$$

$$\frac{\partial P_{lk}}{\partial V_l} = 2 V_l g_{kl} - a_{kl} V_k g_{kl} \cos(\theta_{lk} - \varphi_{kl}) - a_{kl} V_k b_{kl} \sin(\theta_{lk} - \varphi_{kl}). \tag{4.22}$$

- Reactive power flows from bus k to bus l:

$$\frac{\partial Q_{kl}}{\partial \theta_k} = -a_{kl} V_k V_l b_{kl} \sin(\theta_{kl} + \varphi_{kl}) - a_{kl} V_k V_l g_{kl} \cos(\theta_{kl} + \varphi_{kl})$$

$$\frac{\partial Q_{kl}}{\partial \theta_l} = a_{kl} V_k V_l b_{kl} \sin(\theta_{kl} + \varphi_{kl}) + a_{kl} V_k V_l g_{kl} \cos(\theta_{kl} + \varphi_{kl})$$

$$\frac{\partial Q_{kl}}{\partial V_k} = -2a_{kl}^2 V_k\left(b_{kl} + b_{kl}^{sh}\right) + a_{kl} V_l b_{kl} \cos(\theta_{kl} + \varphi_{kl}) - a_{kl} V_l g_{kl} \sin(\theta_{kl} + \varphi_{kl})$$

$$\frac{\partial Q_{kl}}{\partial V_l} = a_{kl} V_k b_{kl} \cos(\theta_{kl} + \varphi_{kl}) - a_{kl} V_k g_{kl} \sin(\theta_{kl} + \varphi_{kl}). \tag{4.23}$$

- Reactive power flows from bus l to bus k:

$$\frac{\partial Q_{lk}}{\partial \theta_k} = a_{kl} V_k V_l b_{kl} \sin(\theta_{lk} - \varphi_{kl}) + a_{kl} V_k V_l g_{kl} \cos(\theta_{lk} - \varphi_{kl})$$

$$\frac{\partial Q_{lk}}{\partial \theta_l} = -a_{kl} V_k V_l b_{kl} \sin(\theta_{lk} - \varphi_{kl}) - a_{kl} V_k V_l g_{kl} \cos(\theta_{lk} - \varphi_{kl})$$

$$\frac{\partial Q_{lk}}{\partial V_k} = a_{kl} V_l b_{kl} \cos(\theta_{lk} - \varphi_{kl}) - a_{kl} V_l g_{kl} \sin(\theta_{lk} - \varphi_{kl})$$

$$\frac{\partial Q_{lk}}{\partial V_l} = -2 V_l\left(b_{kl} + b_{kl}^{sh}\right) + a_{kl} V_k b_{kl} \cos(\theta_{lk} - \varphi_{kl}) - a_{kl} V_k g_{kl} \sin(\theta_{lk} - \varphi_{kl}). \tag{4.24}$$

- Active power injections in bus k:

$$\frac{\partial P_k}{\partial \theta_k} = \sum_{l \in \Omega_k} a_{kl} V_k V_l [g_{kl} \sin(\theta_{kl} + \varphi_{kl}) - b_{kl} \cos(\theta_{kl} + \varphi_{kl})]$$

$$\frac{\partial P_k}{\partial \theta_l} = a_{kl} V_k V_l [-g_{kl} \sin(\theta_{kl} + \varphi_{kl}) + b_{kl} \cos(\theta_{kl} + \varphi_{kl})]$$

$$\frac{\partial P_k}{\partial V_k} = \sum_{l \in \Omega_k} 2a_{kl}^2 V_k g_{kl} - a_{kl} V_l [g_{kl} \cos(\theta_{kl} + \varphi_{kl}) + b_{kl} \sin(\theta_{kl} + \varphi_{kl})]$$

$$\frac{\partial P_k}{\partial V_l} = -a_{kl} V_k [g_{kl} \cos(\theta_{kl} + \varphi_{kl}) + b_{kl} \sin(\theta_{kl} + \varphi_{kl})]. \tag{4.25}$$

- Reactive power injections in bus k:

$$\frac{\partial Q_k}{\partial \theta_k} = \sum_{l \in \Omega_k} -a_{kl} V_k V_l [b_{kl} \sin(\theta_{kl} + \varphi_{kl}) + g_{kl} \cos(\theta_{kl} + \varphi_{kl})]$$

$$\frac{\partial Q_k}{\partial \theta_l} = a_{kl} V_k V_l [b_{kl} \sin(\theta_{kl} + \varphi_{kl}) + g_{kl} \cos(\theta_{kl} + \varphi_{kl})]$$

$$\frac{\partial Q_k}{\partial V_k} = -2V_k b_k^{sh} + \sum_{l \in \Omega_k} \left\{ -2V_k \left(a_{kl}^2 + b_{kl} + b_{kl}^{sh} \right) + V_l [a_{kl} b_{kl} \cos(\theta_{kl} + \varphi_{kl}) - a_{kl} g_{kl} \sin(\theta_{kl} + \varphi_{kl})] \right\}$$

$$\frac{\partial Q_k}{\partial V_l} = a_{kl} V_k [b_{kl} \cos(\theta_{kl} + \varphi_{kl}) - g_{kl} \sin(\theta_{kl} + \varphi_{kl})]. \tag{4.26}$$

The partial derivatives with respect to the active and reactive power injection equations can also be expressed as a function of the EPS admittance matrix, as described in the following:

- Active power injections in bus k:

$$\frac{\partial P_k}{\partial \theta_k} = \sum_{l \in \Omega_k} V_k V_l (-G_{kl} \sin\theta_{kl} + B_{kl} \cos\theta_{kl})$$

$$\frac{\partial P_k}{\partial \theta_l} = V_k V_l (G_{kl} \sin\theta_{kl} - B_{kl} \cos\theta_{kl})$$

$$\frac{\partial P_k}{\partial V_k} = 2V_k G_{kk} + \sum_{l \in \Omega_k} V_l (G_{kl}\cos\theta_{kl} + B_{kl}\sin\theta_{kl})$$

$$\frac{\partial P_k}{\partial V_l} = V_k (G_{kl}\cos\theta_{kl} + B_{kl}\sin\theta_{kl}). \tag{4.27}$$

- Reactive power injections in bus k:

$$\frac{\partial Q_k}{\partial \theta_k} = \sum_{l \in \Omega_k} V_k V_l (G_{kl}\cos\theta_{kl} + B_{kl}\sin\theta_{kl})$$

$$\frac{\partial Q_k}{\partial \theta_l} = -V_k V_l (G_{kl}\cos\theta_{kl} + B_{kl}\sin\theta_{kl})$$

$$\frac{\partial Q_k}{\partial V_k} = -2V_k B_{kk} + \sum_{l \in \Omega_k} V_l (G_{kl}\sin\theta_{kl} - B_{kl}\cos\theta_{kl})$$

$$\frac{\partial Q_k}{\partial V_l} = V_k (G_{kl}\sin\theta_{kl} - B_{kl}\cos\theta_{kl}). \tag{4.28}$$

Note 4.1: Note that the expression $\sum_{l \in \Omega_k}$ refers to all the buses adjacent to the bus k, that is, it also incorporates elements outside the main diagonal.

Finally, because the voltage meters directly measure magnitude voltage, its expression with respect to state of voltage magnitude (v) is direct:

$$v_k = V_k. \tag{4.29}$$

- Voltage magnitudes in bus k:

$$\frac{\partial v_k}{\partial \theta_k} = 0$$

$$\frac{\partial v_k}{\partial \theta_l} = 0$$

$$\frac{\partial v_k}{\partial V_k} = 1$$

$$\frac{\partial v_k}{\partial V_l} = 0. \tag{4.30}$$

Note 4.2: In state estimation, the reference bus is used only as an angular reference, unlike the load flow problem, for which the reference bus also provides the power balance of the system. In the estimation problem, one can choose any bus as an angular reference and, to solve the problem, two paths are usually required: (i) the column of H referring to the reference angle is eliminated from the estimation process; or (ii) this column is maintained by inserting a pseudo measurement of voltage angle for that bus in the measurement vector (assigning the zero value for that measurement).

Example 4.1

For the three-bus radial system associated with the set of measurements presented in Fig. 4.2, whose parameters are presented in Table 4.1, obtain the state variables estimated through the coupled WLS estimator. For this, consider the values of the measurements and respective standard deviations presented in Table 4.2.

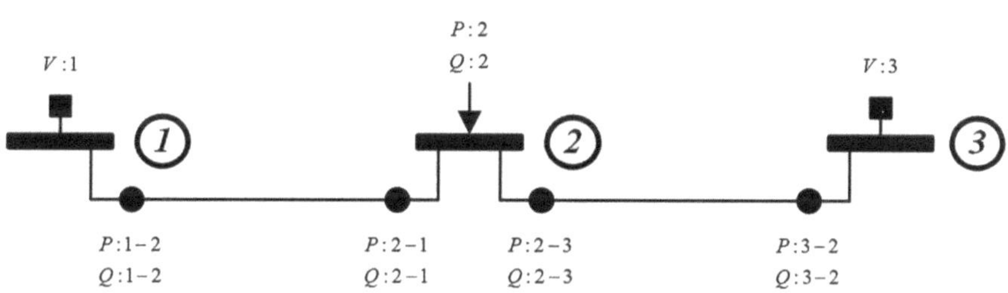

Fig. 4.2 Three-bus power system (with radial structure).

Table 4.1 Parameters of the radial system of three buses.

Line	g (pu)	b (pu)	b^{sh} (pu)
1–2	4.99913	−15.26308	0.02640
2–3	1.13502	−4.78186	0.02190

Table 4.2 Measurement values and respective standard deviations.

Measurement	Value (pu)	Standard deviation (σ)
P:1-2	0.8190	0.005460
P:2-1	−0.8070	0.005380
P:2-3	0.9900	0.006600

Continued

Table 4.2 Measurement values and respective standard deviations—cont'd

Measurement	Value (pu)	Standard deviation (σ)
P:3-2	−0.9420	0.006280
Q:1-2	−0.0110	0.000073
Q:2-1	−0.0120	0.000080
Q:2-3	0.3490	0.002327
Q:3-2	−0.1900	0.001267
P:2	0.1830	0.001220
Q:2	0.3380	0.002253
V:1	1.0600	0.007067
V:3	0.9450	0.006300

Solution: The measurement model is given by

$$
\begin{aligned}
P_{12} &= V_1^2 g_{12} - V_1 V_2 g_{12} \cos(\theta_{12}) - V_1 V_2 b_{12} sen(\theta_{12}) + e_{P12} \\
P_{21} &= V_2^2 g_{21} - V_2 V_1 g_{21} \cos(\theta_{21}) - V_2 V_1 b_{21} sen(\theta_{21}) + e_{P21} \\
P_{23} &= V_2^2 g_{23} - V_2 V_3 g_{23} \cos(\theta_{23}) - V_2 V_3 b_{23} sen(\theta_{23}) + e_{P23} \\
P_{32} &= V_3^2 g_{32} - V_3 V_2 g_{32} \cos(\theta_{32}) - V_3 V_2 b_{32} sen(\theta_{32}) + e_{P32} \\
Q_{12} &= -V_1^2 \left(b_{12} + b_{12}^{sh}\right) + V_1 V_2 b_{12} \cos(\theta_{12}) - V_1 V_2 g_{12} sen(\theta_{12}) + e_{Q12} \\
Q_{21} &= -V_2^2 \left(b_{21} + b_{21}^{sh}\right) + V_2 V_1 b_{21} \cos(\theta_{21}) - V_2 V_1 g_{21} sen(\theta_{21}) + e_{Q21} \\
Q_{23} &= -V_2^2 \left(b_{23} + b_{23}^{sh}\right) + V_2 V_3 b_{23} \cos(\theta_{23}) - V_2 V_3 g_{23} sen(\theta_{23}) + e_{Q23} \\
Q_{32} &= -V_3^2 \left(b_{32} + b_{32}^{sh}\right) + V_3 V_2 b_{32} \cos(\theta_{32}) - V_3 V_2 g_{32} sen(\theta_{32}) + e_{Q32} \\
P_2 &= V_2 \sum_{l \in \Omega_k} V_l \left[G_{2l} \cos(\theta_{2l}) + B_{2l} sen(\theta_{2l})\right] + e_{P2} \\
Q_2 &= V_2 \sum_{l \in \Omega_k} V_l \left[G_{2l} sen(\theta_{2l}) - B_{2l} \cos(\theta_{2l})\right] + e_{Q2} \\
v_1 &= V_1 + e_{v1} \\
v_3 &= V_3 + e_{v3}
\end{aligned}
\tag{4.31}
$$

The system model is given by

$$
\begin{aligned}
y_{kl} &= g_{kl} + j b_{kl}, \ \forall k \neq l \\
y_{kk} &= j b_k^{sh} \\
Y_{kl} &= -y_{kl} \\
Y_{kk} &= y_{kk} + \sum_{l \in \Omega_k} \left(y_{kl} + j b_{kl}^{sh}\right)
\end{aligned}
$$

$$
G = real(Y) = \begin{bmatrix} 4.9991 & -4.9991 & 0 \\ -4.9991 & 6.1341 & -1.1350 \\ 0 & -1.1350 & 1.1350 \end{bmatrix}
$$

$$B = imag(Y) = \begin{bmatrix} -15.2367 & 15.2631 & 0 \\ 15.2631 & -19.9966 & 4.7819 \\ 0 & 4.7819 & -4.7600 \end{bmatrix} \tag{4.32}$$

The weight matrix is given by

$$R^{-1} = \text{diag}\left[\sigma_{P12}^{-2}\ \sigma_{P21}^{-2}\ \sigma_{P23}^{-2}\ \sigma_{P32}^{-2}\ \sigma_{Q12}^{-2}\ \sigma_{Q21}^{-2}\ \sigma_{Q23}^{-2}\ \sigma_{Q32}^{-2}\ \sigma_{P2}^{-2}\ \sigma_{Q2}^{-2}\ \sigma_{V1}^{-2}\ \sigma_{V3}^{-2}\right]. \tag{4.33}$$

The state variables are estimated using the WLS estimator, following the algorithm:

1. Initialize the state variables x^0 in the flat start, that is, nodal voltages with unit values (1 pu) and phase angles of nodal voltages with null values (0 rad).
2. Calculate the matrices $H(x^\kappa)$ and $G(x^\kappa)$ at the point $x^\kappa = x^0$. The Jacobian matrix has the following structure (note that the column corresponding to the state variable θ_1 has been eliminated because the bus 1 has been chosen as the angular reference):

$$H(x) = \begin{bmatrix}
\frac{\partial P_{12}}{\partial \theta_2} & \frac{\partial P_{12}}{\partial \theta_3} & \frac{\partial P_{12}}{\partial V_1} & \frac{\partial P_{12}}{\partial V_2} & \frac{\partial P_{12}}{\partial V_3} \\
\frac{\partial P_{21}}{\partial \theta_2} & \frac{\partial P_{21}}{\partial \theta_3} & \frac{\partial P_{21}}{\partial V_1} & \frac{\partial P_{21}}{\partial V_2} & \frac{\partial P_{21}}{\partial V_3} \\
\frac{\partial P_{23}}{\partial \theta_2} & \frac{\partial P_{23}}{\partial \theta_3} & \frac{\partial P_{23}}{\partial V_1} & \frac{\partial P_{23}}{\partial V_2} & \frac{\partial P_{23}}{\partial V_3} \\
\frac{\partial P_{32}}{\partial \theta_2} & \frac{\partial P_{32}}{\partial \theta_3} & \frac{\partial P_{32}}{\partial V_1} & \frac{\partial P_{32}}{\partial V_2} & \frac{\partial P_{32}}{\partial V_3} \\
\frac{\partial Q_{12}}{\partial \theta_2} & \frac{\partial Q_{12}}{\partial \theta_3} & \frac{\partial Q_{12}}{\partial V_1} & \frac{\partial Q_{12}}{\partial V_2} & \frac{\partial Q_{12}}{\partial V_3} \\
\frac{\partial Q_{21}}{\partial \theta_2} & \frac{\partial Q_{21}}{\partial \theta_3} & \frac{\partial Q_{21}}{\partial V_1} & \frac{\partial Q_{21}}{\partial V_2} & \frac{\partial Q_{21}}{\partial V_3} \\
\frac{\partial Q_{23}}{\partial \theta_2} & \frac{\partial Q_{23}}{\partial \theta_3} & \frac{\partial Q_{23}}{\partial V_1} & \frac{\partial Q_{23}}{\partial V_2} & \frac{\partial Q_{23}}{\partial V_3} \\
\frac{\partial Q_{32}}{\partial \theta_2} & \frac{\partial Q_{32}}{\partial \theta_3} & \frac{\partial Q_{32}}{\partial V_1} & \frac{\partial Q_{32}}{\partial V_2} & \frac{\partial Q_{32}}{\partial V_3} \\
\frac{\partial P_{2}}{\partial \theta_2} & \frac{\partial P_{2}}{\partial \theta_3} & \frac{\partial P_{2}}{\partial V_1} & \frac{\partial P_{2}}{\partial V_2} & \frac{\partial P_{2}}{\partial V_3} \\
\frac{\partial Q_{2}}{\partial \theta_2} & \frac{\partial Q_{2}}{\partial \theta_3} & \frac{\partial Q_{2}}{\partial V_1} & \frac{\partial Q_{2}}{\partial V_2} & \frac{\partial Q_{2}}{\partial V_3} \\
\frac{\partial V_{1}}{\partial \theta_2} & \frac{\partial V_{1}}{\partial \theta_3} & \frac{\partial V_{1}}{\partial V_1} & \frac{\partial V_{1}}{\partial V_2} & \frac{\partial V_{1}}{\partial V_3} \\
\frac{\partial V_{3}}{\partial \theta_2} & \frac{\partial V_{3}}{\partial \theta_3} & \frac{\partial V_{3}}{\partial V_1} & \frac{\partial V_{3}}{\partial V_2} & \frac{\partial V_{3}}{\partial V_3}
\end{bmatrix}. \tag{4.34}$$

The gain matrix, at iteration κ, is given by

$$G(x^\kappa) = H(x^\kappa)^T R^{-1} H(x^\kappa). \tag{4.35}$$

1. Correction is obtained in the state variables through the normal equation and then their values are updated:

$$\Delta x^\kappa = G(x^\kappa)^{-1} H(x^\kappa)^T R^{-1} [z - h(x^\kappa)]$$

$$x^{\kappa+1} = x^\kappa + \Delta x^\kappa \tag{4.36}$$

2. Convergence test:

$$\max|\Delta x^\kappa| \leq \varepsilon. \tag{4.37}$$

If the above condition is satisfied, the process has converged. Otherwise, proceed to the next iteration ($\kappa = \kappa + 1$) and return to Step 2.

Considering bus 1 as the angular reference ($\theta_1 = 0$) and the tolerance $\varepsilon = 0.001$, the result obtained for this example was

$$\hat{x} = \begin{bmatrix} \hat{\theta}_2 \\ \hat{\theta}_3 \\ \hat{V}_1 \\ \hat{V}_2 \\ \hat{V}_3 \end{bmatrix} = \begin{bmatrix} -0.0435 \\ -0.2255 \\ 1.0591 \\ 1.0441 \\ 0.9442 \end{bmatrix}. \tag{4.38}$$

4.3.3 Fast decoupled state estimator

The largest computational effort involving the previously presented WLS estimator algorithm is associated with the calculation of the G matrix and, later, to its factorization at each iteration. As in the load flow problem, an alternative to reduce the calculations involved in the state estimation process is to keep constant the Jacobian matrix that relates the equations of the measurements with the state variables.

For this reason, based on the principle that the active powers are influenced predominantly by the voltage phase angles and that the reactive powers, in turn, by the voltage magnitudes, several versions of decoupled state estimators have been proposed in the literature (Horisberger et al., 1976). Among these versions, we highlight the one developed by Monticelli and Garcia (1990), known as the fast decoupled WLS state estimator, which can obtain good results even in systems with low reactance/resistance ratio. Its formulation will be presented below.

Consider the following decoupled formulation for the system of nonlinear equations that represents the state estimation problem in EPS:

$$z_p = h_p(x) + e_p$$
$$z_q = h_q(x) + e_q, \tag{4.39}$$

where p is the sub index that represents the set of active measurements formed of active power injection and power flow measurements; q is the sub index that represents the set of reactive measurements formed of reactive power injection and power flow measurements, besides voltage magnitude measurements.

The vector of state variables x can also be written as follows:

$$x = \begin{bmatrix} \theta \\ V \end{bmatrix} \tag{4.40}$$

where V is the vector of voltage magnitudes ($n \times 1$) and θ is the vector of voltage angles (($n - 1) \times 1$), where n is the number of buses in the system.

From the division of the set of measurements and the decomposition of the vector of state variables, we can decompose the Jacobian matrix as follows:

$$H(x) = \frac{\partial h(x)}{\partial x} = \begin{bmatrix} \dfrac{\partial h_p(x)}{\partial \theta} & \dfrac{\partial h_p(x)}{\partial V} \\ \dfrac{\partial h_q(x)}{\partial \theta} & \dfrac{\partial h_q(x)}{\partial V} \end{bmatrix} \tag{4.41}$$

Through the fast decoupled WLS estimator (Monticelli and Garcia, 1990), the solution of the problem is obtained through an iterative procedure, whose corrections $\Delta\theta$ and ΔV are calculated separately. In this approach, the effects of the submatrices H_{pV} and $H_{q\theta}$ are neglected and the submatrices $H_{p\theta}$ and H_{qV} are kept constant during the iterative process. Note that the introduction of approximations in the Jacobian matrix alters the convergence process, that is, it changes the path traveled between the starting point and the solution, but does not change the final solution, since the problem to be solved remains the same (Monticelli, 1999).

The submatrices $H_{p\theta}$ and H_{qV} (considered fixed along the iterations) are calculated according to the flat start ($\theta = 0$ *rad* and $V = 1$ pu) and neglecting the series resistors in the formation of submatrix H_{qV}. From these considerations, the optimal solution by weighted least squares, fast decoupled version, can be obtained through two steps (Monticelli and Garcia, 1990):

(i) Calculation of the active iteration $P\theta$:

$$\Delta\theta^{\kappa} = \left(H_{p\theta}^{T} R_{p}^{-1} H_{p\theta}\right)^{-1} H_{p\theta}^{T} R_{p}^{-1}\left[z_{p} - h_{p}(\theta^{\kappa}, V^{\kappa})\right],$$

$$\theta^{\kappa+1} = \theta^{\kappa} + \Delta\theta^{\kappa}. \tag{4.42}$$

(ii) Calculation of the reactive iteration QV:

$$\Delta V^{\kappa} = \left(H_{qV}^{T} R_{q}^{-1} H_{qV}\right)^{-1} H_{qV}^{T} R_{q}^{-1}\left[z_{q} - h_{q}\left(\theta^{\kappa+1}, V^{\kappa}\right)\right],$$

$$V^{\kappa+1} = V^{\kappa} + \Delta V^{\kappa}. \tag{4.43}$$

The iterative process is finalized when the magnitudes of the corrections $\Delta\theta$ and ΔV are smaller than the adopted stop criteria, for example:

$$\max\left|\Delta\theta^{\kappa}\right| \leq \varepsilon_{p\theta}$$
$$\max\left|\Delta V^{\kappa}\right| \leq \varepsilon_{qV}, \tag{4.44}$$

in which $\varepsilon_{p\theta}$ and ε_{qV} denote, respectively, the tolerance to the approximation error of the active model and the reactive model. In addition, when one of the stopping criteria is reached, then the respective state vector is fixed and only the half iteration whose tolerance has not yet been obtained is processed and, consequently, the convergence process becomes even faster.

4.3.4 Linear WLS state estimator

The linear WLS state estimator, although seldom applied in practice, is of considerable importance in studies of methods and techniques for state estimation. This is because the simplifying assumptions on which it is based transform the estimation problem into a noniterative form, making it easy to understand issues related to observability and redundancy of measurements.

The linearized estimator is based on linear load flow equations, where losses are neglected, and all voltage magnitudes are 1 pu. As a function of these approximations, the active power flow between buses k and l, as well as the active power injection in bus k, are calculated, respectively, by the following expressions:

$$P_{kl} = \frac{\theta_k - \theta_l}{x_{kl}}, \tag{4.45}$$

$$P_k = \sum_{l \in \Omega_k} P_{kl}, \tag{4.46}$$

where θ_k and θ_l are, respectively, the voltage phase angles in the buses k and l, x_{kl} is the reactance of the transmission line connecting the buses k and l, and Ω_k is the set of buses connected to the bus k.

The active power flow equation written in this way allows analogies with the Ohm's law for a direct current (DC) circuit. Because of this, the linearized state estimator is known as the DC estimator.

In the linear formulation, the measurement model presented in Eq. (4.2) becomes (Monticelli, 1999):

$$z_{ap} = H_{DC}\theta + e_{ap} \tag{4.47}$$

where z_{ap} is the vector of active power measurements ($m_{ap} \times 1$); H_{DC} is the Jacobian matrix of the linear estimator ($m_{ap} \times N_{ap}$); θ is the vector of state variables ($N_{ap} \times 1$) which, in the linear formulation, corresponds to the voltage phase angles in the system buses; e_{ap} is the vector of errors of the active power measurements ($m_{ap} \times 1$), which is assumed to have Gaussian distribution with zero mean and covariance matrix R_{ap}; m_{ap} is the number of active power measurements; N_{ap} is the number of state variables to be estimated ($N_{ap} = n - 1$).

The estimate of the state vector θ, designated by $\hat{\theta}$, obtained by the weighted least squares method, is the value of θ that makes the index $J(\theta)$ minimum, given by

$$J(\theta) = e_{ap}^T R_{ap}^{-1} e_{ap} = \left[z_{ap} - H_{DC}\theta\right]^T R_{ap}^{-1} \left[z_{ap} - H_{DC}\theta\right]. \tag{4.48}$$

Thus the optimal solution can be calculated by

$$\hat{\theta} = \left[H_{DC}^T R_{ap}^{-1} H_{DC}\right]^{-1} H_{DC}^T R_{ap}^{-1} z_{ap}. \tag{4.49}$$

Example 4.2

Consider the electric power system illustrated in Fig. 4.3. The values observed by the meters presented in this figure, with the respective standard deviations, are presented in Table 4.3.

Using the least-squares linear static state estimator:

(a) Estimate the state variables of the system considering as weight matrix the identity matrix.

(b) Estimate the system state variables by applying the weighted version of the estimator, using the standard deviations shown in Table 4.3.

Solution:

Analyzing the system in question, we have the following data:

Lines reactances:

$$x_{12} = x_{21} = 0.04\,pu$$

$$x_{13} = x_{31} = 0.05\,pu$$

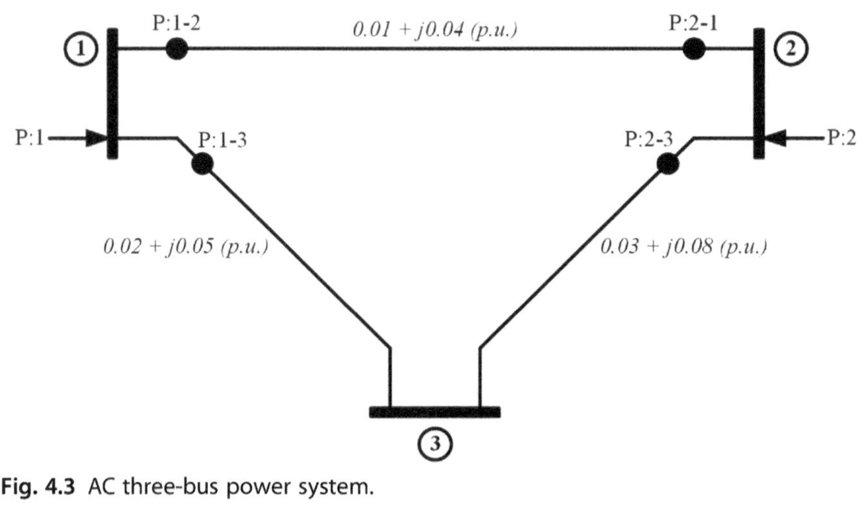

Fig. 4.3 AC three-bus power system.

Table 4.3 Measurements of active power and respective standard deviations.

Measurement	Value (pu)	Standard deviation (σ)
P:1-2	6.2452	0.008
P:2-1	−6.2340	0.008
P:1-3	6.9920	0.008
P:2-3	1.1300	0.008
P:1	13.2650	0.010
P:2	−5.0055	0.010

$$x_{23} = x_{32} = 0.08\,pu$$

Measurement model:

$$P_{12} = \frac{\theta_1 - \theta_2}{x_{12}} + e_{P12}$$

$$P_{21} = \frac{\theta_2 - \theta_1}{x_{21}} + e_{P21}$$

$$P_{13} = \frac{\theta_1 - \theta_3}{x_{13}} + e_{P13}$$

$$P_{23} = \frac{\theta_2 - \theta_3}{x_{23}} + e_{P23}$$

$$\begin{aligned} P_1 = P_{12} + P_{13} &= \frac{\theta_1 - \theta_2}{x_{12}} + \frac{\theta_1 - \theta_3}{x_{13}} + e_{P1} \\ &= \frac{(x_{12} + x_{13})\theta_1 - x_{13}\theta_2 - x_{12}\theta_3}{x_{12}x_{13}} + e_{P1} \end{aligned}$$

$$\begin{aligned} P_2 = P_{21} + P_{23} &= \frac{\theta_2 - \theta_1}{x_{21}} + \frac{\theta_2 - \theta_3}{x_{23}} + e_{P2} \\ &= \frac{-x_{23}\theta_1 + (x_{21} + x_{23})\theta_2 - x_{21}\theta_3}{x_{21}x_{23}} + e_{P2} \end{aligned}$$

Matrix form:

$$\begin{bmatrix} P_{12} \\ P_{21} \\ P_{13} \\ P_{23} \\ P_1 \\ P_2 \end{bmatrix} = \begin{bmatrix} \frac{1}{x_{12}} & \frac{-1}{x_{12}} & 0 \\ \frac{-1}{x_{21}} & \frac{1}{x_{21}} & 0 \\ \frac{1}{x_{13}} & 0 & \frac{-1}{x_{13}} \\ 0 & \frac{1}{x_{23}} & \frac{-1}{x_{23}} \\ \frac{x_{12} + x_{13}}{x_{12}x_{13}} & \frac{-1}{x_{12}} & \frac{-1}{x_{13}} \\ \frac{-1}{x_{21}} & \frac{x_{21} + x_{23}}{x_{21}x_{23}} & \frac{-1}{x_{23}} \end{bmatrix} \begin{bmatrix} \theta_1 \\ \theta_2 \\ \theta_3 \end{bmatrix} + \begin{bmatrix} e_{P12} \\ e_{P21} \\ e_{P13} \\ e_{P23} \\ e_{P1} \\ e_{P2} \end{bmatrix}$$

Replacing the values of the reactances, we have

$$\underbrace{\begin{bmatrix} P_{12} \\ P_{21} \\ P_{13} \\ P_{23} \\ P_1 \\ P_2 \end{bmatrix}}_{} = \underbrace{\begin{bmatrix} 25.0 & -25.0 & 0 \\ -25.0 & 25.0 & 0 \\ 20.0 & 0 & -20.0 \\ 0 & 12.5 & -12.5 \\ 45.0 & -25.0 & -20.0 \\ -25.0 & 37.5 & -12.5 \end{bmatrix}}_{} \underbrace{\begin{bmatrix} \theta_1 \\ \theta_2 \\ \theta_3 \end{bmatrix}}_{} + \underbrace{\begin{bmatrix} e_{P12} \\ e_{P21} \\ e_{P13} \\ e_{P23} \\ e_{P1} \\ e_{P2} \end{bmatrix}}_{}$$

Choosing bus 1 as reference ($\theta_1 = 0$ *rad*), the above system becomes:

$$\underbrace{\begin{bmatrix} P_{12} \\ P_{21} \\ P_{13} \\ P_{23} \\ P_1 \\ P_2 \end{bmatrix}}_{z_{ap}} = \underbrace{\begin{bmatrix} -25.0 & 0 \\ 25.0 & 0 \\ 0 & -20.0 \\ 12.5 & -12.5 \\ -25.0 & -20.0 \\ 37.5 & -12.5 \end{bmatrix}}_{H_{DC}} \underbrace{\begin{bmatrix} \theta_2 \\ \theta_3 \end{bmatrix}}_{\theta} + \underbrace{\begin{bmatrix} e_{P12} \\ e_{P21} \\ e_{P13} \\ e_{P23} \\ e_{P1} \\ e_{P2} \end{bmatrix}}_{e_{ap}}$$

(a) State variables estimated using the WLS linear estimator considering, as the weight matrix, the identity matrix:

$$\hat{\theta} = \left(H_{DC}^T H_{DC}\right)^{-1} H_{DC}^T z_{ap} = \begin{bmatrix} -0.2504 \\ -0.3488 \end{bmatrix}$$

That is, $\theta_2 = -0.2504$ rad and $\theta_3 = -0.3488$ rad.

(b) The weight matrix is given by

$$R_{ap}^{-1} = \text{diag}\left[\sigma_{P12}^{-2}\ \sigma_{P21}^{-2}\ \sigma_{P13}^{-2}\ \sigma_{P23}^{-2}\ \sigma_{P1}^{-2}\ \sigma_{P2}^{-2}\right].$$

State variables estimated using the WLS linear method:

$$\hat{\theta} = \left(H_{DC}^T R_{ap}^{-1} H_{DC}\right)^{-1} H_{DC}^T R_{ap}^{-1} z_{ap} = \begin{bmatrix} -0.2504 \\ -0.3484 \end{bmatrix}.$$

That is, $\theta_2 = -0.2504$ rad and $\theta_3 = -0.3484$ rad.

4.4 Alternative WLS estimator formulations

Because the WLS estimator based on the normal equation (Eq. 4.15) is prone to poor numerical conditioning, more numerically robust alternative formulations have been proposed over the years. To understand this trend, it is important to analyze the solution process of the estimation problem from the normal equation, which requires the obtaining and factorization of the matrix $G(x^\kappa)$.

The matrix $G(x^\kappa)$ is obtained through the matrix product $H(x^\kappa)^T R^{-1} H(x^\kappa)$. As the matrix R^{-1} is diagonal, the matrix $G(x^\kappa)$ is symmetric in structure and value, and positive definite for observable systems, which facilitates its factorization and, as a consequence, the solution of the linear system presented by the normal equation. Among the factorization methods traditionally used in state estimation, we can mention Gauss elimination method, LDU factorization, Cholesky factorization, and QR decomposition, among others presented in Golub and van Loan (1996).

The numerical robustness of the solution of the normal equation will depend not only on the factorization method used, but also on the numerical conditioning of the Gain matrix (Ebrahimian and Baldick, 2001).

The condition number of a nonsingular matrix A, represented by $K(A)$, can be defined by

$$K(A) = \|A\|\left\|A^{-1}\right\|, \tag{4.50}$$

where $\|.\|$ represents any matrix norm.

A high-condition number results in a high sensitivity of the linear system solution in relation to changes in the input data. In addition, it indicates that the matrix under analysis is poorly conditioned. For the case of a singular matrix A, $K(A)$ tends to infinity. Thus it can be said that the greater the condition number of a matrix, the closer it will be to the singularity.

In Quarteroni et al. (2000), it is proved that the condition number of a matrix, calculated through the Euclidean norm, is reduced to

$$K_2(A) = \|A\|_2\left\|A^{-1}\right\|_2 = \frac{\sigma_{max}(A)}{\sigma_{min}(A)}, \tag{4.51}$$

where σ_{max} and σ_{min} represent, respectively, the maximum and minimum singular value of A.

For the case where a matrix B ($B = A^TA$) is symmetric and positive definite, we have

$$K_2(B) = \frac{\lambda_{max}(B)}{\lambda_{min}(B)} = \frac{\lambda_{max}(A^TA)}{\lambda_{min}(A^TA)} = \frac{\sigma^2_{max}(A)}{\sigma^2_{min}(A)} = [K_2(A)]^2, \tag{4.52}$$

where λ_{max} is the largest eigenvalue and λ_{min} is the smallest eigenvalue of B.

From these results and considering R^{-1} equal to the identity matrix, the condition number of the gain matrix of the normal equation can be approximated by

$$K_2(G) = \frac{\lambda_{max}(G)}{\lambda_{min}(G)} \cong [K_2(H)]^2. \tag{4.53}$$

Thus it is concluded that the method of the normal equation is poorly conditioned, since the H^TH product in the calculation of the matrix G causes the gain matrix to have worse conditioning than H, in a quadratic order of magnitude (Golub and van Loan, 1996). In the literature, research points out some situations in which the problem of numerical conditioning of the G matrix can be aggravated (Holten et al., 1988; Ebrahimian and Baldick, 2001; Abur and Expósito, 2004; Hebling et al., 2020):

1. Disparity between weights of measurements (elements of R^{-1}), a problem that is especially the case when virtual measurements (zero-injection buses) are treated as measurements with low variance.
2. Injection measurements on buses connected to several transmission lines.
3. Large number of injection measurements.
4. The existence of very short and very long transmission lines in the same system.

In the following subsections will be presented formulations derived from the traditional WLS method that aim to reduce the number of calculations needed and information stored in computer, substantially improving the numerical precision of the solution.

4.4.1 WLS estimator with equality constraints

As previously mentioned, commonly in an EPS there are buses whose injections of active and reactive power are zero (transfer buses). The information associated with these buses is also used in the PSSE process, defined as virtual measurements, which receive a special treatment, since, strictly speaking, they are correct.

Traditionally, virtual measurements are modeled in two ways (Abur and Expósito, 2004):

1. As high-precision power injection measurements (low variance), assigning them high weights. Although it is a simple procedure, the presence of very discrepant weights in the same set of measurements can contribute to the increase in the condition number of the gain matrix and to numerically impair the convergence of the WLS estimator based on the normal equation.
2. Another way of treating zero injections is to represent them as equality constraints on the weighted least squares problem, in order to reduce numerical problems. Thus the estimation problem becomes

$$\text{minimize}: J(x) = [z - h(x)]^T R^{-1} [z - h(x)]$$

$$\textit{subject to}: c(x) = 0, \tag{4.54}$$

where $c(x) = 0$ represents the vector of nonlinear equations of zero power injections (virtual measurements).

Consequently, the solution of the problem, represented by Eq. (4.54), is formulated through the Lagrangian function:

$$\mathcal{L}(x, \lambda) = [z - h(x)]^T R^{-1} [z - h(x)] - \lambda^T c(x), \tag{4.55}$$

in which λ is the vector of Lagrange multipliers, related to the equality constraints, whose dimension is equal to the number of zero injections.

Applying to Eq. (4.55), the optimality conditions of the first order, we obtain the following set of equations:

$$\frac{\partial \mathcal{L}(x, \lambda)}{\partial x} = 0 \Rightarrow H^T(x) R^{-1} [z - h(x)] + C^T(x)\lambda = 0$$

$$\frac{\partial \mathcal{L}(x, \lambda)}{\partial \lambda} = 0 \Rightarrow c(x) = 0, \tag{4.56}$$

where $C(x) = \partial c(x)/\partial x$ is the matrix of first derivatives of the nonlinear functions of the vector $c(x)$ with respect to the vector of state variables x.

Since the set of Eq. (4.56) includes nonlinear functions, it requires the iterative solution of a linear system, with the following form:

$$H^T(x^\kappa) R^{-1} [z - h(x^\kappa) - H(x^\kappa)\Delta x^\kappa] + C^T(x^\kappa)\lambda^\kappa = 0$$

$$c(x^\kappa) + C(x^\kappa)\Delta x = 0, \tag{4.57}$$

which can be represented in matrix form as

$$\begin{bmatrix} H^T(x^\kappa) R^{-1} H(x^\kappa) & C^T(x^\kappa) \\ C(x^\kappa) & 0 \end{bmatrix} \begin{bmatrix} \Delta x^\kappa \\ -\lambda^\kappa \end{bmatrix} = \begin{bmatrix} H^T(x^\kappa) R^{-1} [z - h(x^\kappa)] \\ -c(x^\kappa) \end{bmatrix} \tag{4.58}$$

One drawback of this approach is that the matrix of coefficients is indefinite, even for observable systems, differently from what happens in the case of the gain matrix. In order to overcome this problem, we must perform the permutation of lines during the factorization of the system of equations represented by Eq. (4.58). In Gjelsvik et al. (1985), this procedure is modified through the 2 × 2 pivot, which makes the numerical solution more stable when compared to the traditional factorization.

It is also possible to improve the numerical conditioning of the matrix of coefficients through a simple scaling of the Lagrangian function, given by Eq. (4.55), as follows (Abur and Expósito, 2004):

$$\mathcal{L}(x, \lambda) = \alpha [z - H(x)]^T R^{-1} [z - H(x)] - \lambda_s^T c(x), \tag{4.59}$$

where $\lambda_s = \alpha\lambda$ and α is a scalar, generally chosen as

$$\alpha = \frac{1}{\max\{R_{ii}^{-1}\}} \quad or \quad \alpha = \frac{m}{\sum_{i=1}^{m} R_{ii}^{-1}}. \tag{4.60}$$

Consequently, the system of equations to be solved iteratively becomes

$$\begin{bmatrix} \alpha H^T(x^k)R^{-1}H(x^k) & C^T(x^k) \\ C(x^k) & 0 \end{bmatrix} \begin{bmatrix} \Delta x^k \\ -\lambda^k \end{bmatrix} = \begin{bmatrix} \alpha H^T(x^k)R^{-1}[z - h(x^k)] \\ -c(x^k) \end{bmatrix}. \tag{4.61}$$

Although the modeling of zero injections as equality constraints provides a substantial improvement in the numerical conditioning of the coefficient matrix, this does not eliminate the problem of ill conditioning (for example, due to the occurrence of lines of low reactance and measurements with discrepant weights), which considerably affects the numerical stability.

In order to circumvent these limitations, some even more numerically robust methods have emerged. A review of these methods can be found in Holten et al. (1988), Monticelli (2000), and Abur and Expósito (2004). The methods used in this work are the Hachtel matrix (Gjelsvik et al., 1985), the Sparse Tableau method (Nucera and Gilles, 1991), and the orthogonal transformations method (Costa and Quintana, 1981a,b).

4.4.2 Hachtel augmented matrix method

In this method, the residual vector r is considered an unknown and its associated equation is treated as an equality constraint in the formulation of the state estimation problem, as well as performed for zero injections in the method presented in the previous section. From these considerations, the optimization problem becomes

$$\text{minimize}: J(x) = r^T R^{-1} r$$

$$subject\ to: \begin{cases} c(x) = 0 \\ r - z + h(x) = 0 \end{cases} \tag{4.62}$$

where $c(x) = 0$ is the vector of nonlinear equations related to virtual measurements (zero-power injections) and R^{-1} is the weight matrix of the remaining measurements, also known as regular measurements (analog measurements obtained in real-time and eventually pseudo measurements inserted for restoration of observability).

Consequently, the following Lagrangian function is obtained:

$$\mathcal{L}(x, r, \lambda, \mu) = \left[r^T R^{-1} r\right] - \lambda^T c(x) - \mu^T [r - z + h(x)], \tag{4.63}$$

where λ is the vector of Lagrange multipliers related to the virtual measurements and μ is the vector of Lagrange multipliers related to the residual constraints.

From the application of the first-order optimality conditions, we obtain the following set of equations:

$$\begin{aligned}
\frac{\partial \mathcal{L}(x, r, \lambda, \mu)}{\partial x} &= 0 \Rightarrow C^T(x)\lambda + H^T(x)\mu = 0; \\
\frac{\partial \mathcal{L}(x, r, \lambda, \mu)}{\partial r} &= 0 \Rightarrow R^{-1} r - \mu = 0; \\
\frac{\partial \mathcal{L}(x, r, \lambda, \mu)}{\partial \lambda} &= 0 \Rightarrow c(x) = 0; \\
\frac{\partial \mathcal{L}(x, r, \lambda, \mu)}{\partial \mu} &= 0 \Rightarrow r - z + h(x) = 0.
\end{aligned} \tag{4.64}$$

By analyzing the second equation of the set above, one can conclude that $r = R\mu$. This allows the variable r to be eliminated from the equation. Consequently, the iterative solution of the set of Eq. (4.64) can be obtained by the following matrix equation:

$$\begin{bmatrix} R & H(x^\kappa) & 0 \\ H^T(x^\kappa) & 0 & C^T(x^\kappa) \\ 0 & C(x^\kappa) & 0 \end{bmatrix} \begin{bmatrix} \mu^\kappa \\ \Delta x^\kappa \\ \lambda^\kappa \end{bmatrix} = \begin{bmatrix} z - h(x^\kappa) \\ 0 \\ -c(x^\kappa) \end{bmatrix}. \tag{4.65}$$

Like the coefficient matrix of Eq. (4.58), the matrix of coefficients (Hachtel matrix) of Eq. (4.65) is also indefinite. Therefore the application of the 2×2 pivot is also recommended to make the numerical solution more stable (Gjelsvik et al., 1985). Another point to be highlighted is that the condition number of the coefficient matrix is smaller for this modeling, when compared to that presented in Eq. (4.58). In addition, the array condition number can still be reduced through a simple Lagrangian function scaling, as performed in Section 4.4.1.

4.4.3 Sparse tableau method

The Tableau Sparta method can be seen as a particular case of the Hachtel augmented matrix method, because in this approach the set of measurements is divided into two sets: one formed by the power flow measurements and voltage magnitudes (set F) and the other formed by the power injection measurements (set I). In this way, the system of equations becomes (Abur and Expósito, 2004):

$$\begin{bmatrix} R_F & 0 & H_F & 0 \\ 0 & R_I & H_I & 0 \\ H_F^T & H_I^T & 0 & C^T \\ 0 & 0 & C & 0 \end{bmatrix} \begin{bmatrix} \mu_F^\kappa \\ \mu_I^\kappa \\ \Delta x^\kappa \\ \lambda^\kappa \end{bmatrix} = \begin{bmatrix} z_F - h_F(x^\kappa) \\ z_I - h_I(x^\kappa) \\ 0 \\ -c(x^\kappa) \end{bmatrix}. \tag{4.66}$$

This ordering reduces the occurrence of so-called fill-ins, that is, of the nonzero elements that arise during the factorization of the coefficient matrix. This directly influences the accuracy of the solution of the system of equations, since it brings as a benefit the reduction of possible numerical errors from the factorization.

4.4.4 Orthogonal transformation method

Consider the transformed model of Eq. (4.15), or normal equation, represented as follows:

$$\left(\tilde{H}^T \tilde{H}\right)\Delta x = \tilde{H}^T \Delta \tilde{z}, \tag{4.67}$$

where $\tilde{H} = R^{-\frac{1}{2}} H$ and $\Delta \tilde{z} = R^{-\frac{1}{2}}[z - h(x)]$ represent, respectively, the weighted Jacobian matrix and the weighted Δz vector.

Through the QR orthogonal transformation method (Costa and Quintana 1981a,b), the weighted Jacobian matrix $\tilde{H}$ can be decomposed into

$$\tilde{H} = QR_t, \tag{4.68}$$

where Q is an m-by-m orthogonal matrix ($QQ^T = I$ or $Q^T = Q^{-1}$) and R_t is an m x N upper trapezoidal matrix.

By properly ordering Q and R_t, the decomposition given by Eq. (4.68) can be written as

$$\tilde{H} = QR_t = [Q_n \ Q_0]\begin{bmatrix} U \\ 0 \end{bmatrix} = Q_n U, \tag{4.69}$$

where Q_n is an $m \times N$ submatrix given by the first N columns of Q and U is an upper triangular matrix $N \times N$.

Substituting Eq. (4.68) into Eq. (4.67), we have the following development:

$$(QR_t)^T (QR_t)\Delta x = (QR_t)^T \Delta \tilde{z}$$

$$R_t^T Q^T Q R_t \Delta x = R_t^T Q^T \Delta \tilde{z}$$

$$R_t^T R_t \Delta x = R_t^T Q^T \Delta \tilde{z}. \tag{4.70}$$

Now, using Eq. (4.69), we have

$$U^T U \Delta x = U^T Q_n^T \Delta \tilde{z}. \tag{4.71}$$

The function $J(x)$ represented in L-2 norm was taken as the starting point for the formulation of the problem in Costa and Quintana (1981a). Thus Eq. (4.71) is obtained through the so-called Golub method, which is based on successive Householder orthogonal transformations on the system of equations. In Costa and Quintana (1981b), the application of the Givens Rotation method is investigated and compared to the previous one. As a result, the latter has the same robust numerical behavior and requires less processing time than the previous one. Therefore this method becomes more attractive for practical and real-time applications.

4.5 WLS estimator considering SCADA measurements and SPMs

Among the hybrid state estimators based on the simultaneous processing of SCADA and SPMs, we highlight the one proposed in Korres and Manousakis (2011), which uses the WLS estimator and has the following characteristics:

1. It does not require many changes in the conventional formulation of the PSSE process by WLS.
2. It does not require the choice of a bus as an angular reference.
3. Enables the processing of gross errors in SCADA and SPMs, if redundancy is adequate (processing of gross errors will be dealt with in Chapter 6).

As mentioned in Chapter 2, in the conventional PSSE process, i.e., without considering SPMs, one of the system buses is chosen as the angular reference. Thus zero value is assigned to the voltage phase angle of this bus and the estimated angles for the other buses are relative to that bus. Since the direct monitoring of voltage phase angles is not usual via the conventional SCADA system, this practice is appropriate, since the absolute value of the reference for bus voltage angle is irrelevant to the PSSE process (London et al., 2009). However, in the hybrid state estimation process, that is, when conventional measurements and SPMs are presented, this practice is not adequate. This is because the voltage phase angle measurements provided by the PMUs are obtained from another reference frame, which is determined by the instant of time provided by the reception of the signal emitted by the GPS system.

Several solutions have been proposed to overcome the problem of choosing the voltage angle reference in the PSSE process, which makes use of simultaneous processing of conventional measurements and SPMs.

Using the WLS hybrid state estimator proposed in Korres and Manousakis (2011), the problem of choosing the angular reference was solved in a very simple way. In the existence of at least one voltage phase angle measurement, the angle reference will be determined by the signal emitted by the GPS system. However, in the absence of such measurements, conventional practice will be used, i.e., one of the system buses is chosen as the angular reference.

The measurement model and the WLS hybrid state estimator solution algorithm proposed in Korres and Manousakis (2011) are practically the same as the conventional WLS estimator already presented in this chapter. The differences are as follows:

- Inclusion of voltage and current SPMs. As a result, the Jacobian matrix of the hybrid estimator has new rows related to SPMs. Also, in the presence of at least one phasor measurement of voltage angle, this matrix will have one more column associated with this state variable, because in this situation the angle reference will be determined by the signal emitted by the GPS system, making it necessary to estimate the voltage phase angles of all the buses of the EPS.
- It uses the measurement model in rectangular coordinates for voltage and current phasors, not polar coordinates commonly used in the measurement model considering only SCADA measurements. This is to avoid numerical problems in initializing the algorithm in the presence of current phasor measurements, since some derivatives of these measurements (elements of the corresponding Jacobian matrix) are undefined when the first iteration starts from the flat start (magnitudes of nodal voltages equal to 1.0 pu and phase angles equal to zero radians for all EPS buses).

Hybrid estimators based on the simultaneous processing of SCADA measurements and SPMs have presented good results in the literature. However, as discussed in Chapter 2, this cannot be guaranteed in practice, since these are measurements with different precisions and sampling rates (Costa et al., 2013).

4.6 Statistically robust state estimators

As presented in Chapter 1, the WLS state estimator works very well when the errors in the measurements are Gaussian, failing, however, in the occurrence of one or more GEs.

To overcome this limitation, several techniques for processing GEs (detection and identification) based on the analysis of the residuals of the WLS estimator were developed, as will be presented in the next chapter. In addition, state estimators more statistically robust than the WLS estimator were applied to EPS, such as those based on the weighted least median of squares (WLMS) and the weighted least absolute value (WLAV) method, proposed in Irving et al. (1978).

4.6.1 WLMS state estimator

The WLMS estimator presented in Mili et al. (1991) is the first statistically robust estimator applied in EPS that is capable of filtering existing GEs in more than one measurement classified as a leverage point[a] (Mili et al., 1991; Zhao and Mili, 2018).

Afterward, Mili et al. (1994) developed an expression to determine the median that minimizes the square-weighted residual considering the local redundancy of measurements, unlike previous work, where this expression was obtained in terms of global redundancy.

The WLMS estimator works as follows: among the m available measurements, a series of samples is drawn with N measurements, where N is the number of state variables to be estimated, so that the EPS is observable. For each sample drawn, the estimator calculates the state variables x and the set of residuals r, generated from x, for all m measurements. Then, these residuals are weighted, squared, and sorted in an increasing fashion ($r^2_{R(1)} \leq \ldots \leq r^2_{R(m)}$). The objective function to be minimized is the median of the set of squares of the weighted residues, corresponding to each selected set of measurements. At the end of the process, the estimated state $\hat{x}$ corresponds to that sample that generated the set of residuals r that has the least-squared weighted mean residual.

The WLMS state estimator has the following formulation:

[a] Leverage points are highly influential measurements that "attract" the state estimation solution. Consequently, even when these measurements contain GEs, their residuals will be very small (Zhao and Mili, 2018).

$$\text{minimize}\big|_{x_i} \; median\big|_{\nu}\left(r_{R(\nu)}^2\right), \tag{4.72}$$

where x_i is the vector of state variables obtained by the measured N samples, ν is the median position of the weighted squared residuals, and $r_{R(\nu)}^2$ is the median weighted squared residual. In the case of linear regression, the median of the residuals is given by Mili et al. (1991):

$$\nu = \left[\frac{m}{2}\right] + \left[\frac{N+1}{2}\right], \tag{4.73}$$

where [.] denotes the integer part of the argument.

The initial step of the algorithm is to randomize, among the measurements available, a series of k samples with N measurements, so that the electrical system is observable. For each of the samples, the state variables and the weighted residual r_R of the m available measurements are calculated. The equation of the weighted residual for the ith measurement is defined as (Monticelli, 1999):

$$r_{R_i} = \frac{r_i}{\sqrt{R_{ii}}}, \tag{4.74}$$

where R_{ii} is the ith diagonal element of the covariance matrix of the measurements.

Then, these weighted residuals are squared and ordered in an increasing manner, that is

Sample 1: $r_{R(1)}^2 \leq \ldots \leq r_{R(\nu)}^2 \leq \ldots \leq r_{R(m)}^2$
Sample 2: $r_{R(1)}^2 \leq \ldots \leq r_{R(\nu)}^2 \leq \ldots \leq r_{R(m)}^2$
⋮
Sample k: $r_{R(1)}^2 \leq \ldots \leq r_{R(\nu)}^2 \leq \ldots \leq r_{R(m)}^2$.

The WLMS estimate corresponds to that sample which minimizes the νth squared weighted residual. Given the above, the algorithm of the WLMS estimator can be defined by the following steps:

- **Step 1:** Through a combinatorial procedure, k distinct measurement sets are randomized each one with N measurements, so that the system is observable for any of these base sets. To ensure that the system is observable for any of the bases, an observability analysis is performed. In Mili et al. (1991), the method proposed by Wu and Monticelli (1985) was used.
- **Step 2:** Calculate the vector of state variables for one of the base sets by a load flow method.

- **Step 3:** Using the vector of state variables calculated in the previous step, calculate the weighted residual of the m measurements available in the system. Obtain the square of each residual and order them incrementally.
- **Step 4**: Through Eq. (4.73), determine the position ν of the median of the residuals.
- **Step 5:** The WLMS estimate is the state vector calculated from the base that has the smallest νth squared weighted residual.

According to Mili et al. (1991), the WLMS estimator is a statistically robust estimator, that is, with a high breaking point[b].

4.7 State estimators for distribution systems

Historically DSs were operated passively, without monitoring and control (Manitsas et al., 2012). The feeders were, and still are mostly, unidirectional connections between the transmission system and the end consumers. In this way, traditionally the procedure used to determine the operating state of the feeders of a DS is based on the load profiles of these feeders. Because the loads are varied and dispersed in the DSs, accurate determination of the load profile of a feeder is a huge challenge, usually involving the development of standardized load profiles for each type of consumer (residential, commercial, industrial, etc.), based on some kind of load monitoring, such as energy measurement. Since such profiles are only a rough approximation to demand for a given load, the quality of the estimates obtained for the state of operation of a feeder through this procedure is low (Baran, 2012).

In the last decades, however, a paradigm shift has arisen in the operation and design of DSs, mainly due to factors related to the concept of smart grids, such as new electrical network architectures with the capacity to operate autonomously; penetration of new renewable energy sources (in primary and secondary distribution networks); the advent of electric vehicles; etc.

Thus these factors have contributed to increase the complexity of the operation of DSs, which are becoming increasingly complex systems.

In view of the previous, in recent years there has been increasing interest in the development and implementation of tools for real-time monitoring of DSs, that is, tools that allow the real-time assessment of the operational condition of the DSs, since from this information control actions can be taken in real-time to correct failures and ensure a safe operating state.

[b] See Appendix A.

As already pointed out, the fundamental tool for real-time monitoring of electrical networks is the state estimator. However, the limited amount of real-time measurements usually available in DSs made it impossible to implement state estimators in these networks. However, recent events, also related to the concept of smart grids, have led to a significant increase in the number of measurements available in DSs, motivating the development and implantation of state estimators for these systems. Among these events are the installation of smart meters in the secondary network and the development of so-called micro-PMUs (Primadianto and Lu, 2017), which allow the measurement of phasor quantities in DSs.

To develop state estimators for DSs, some research has made direct use of methods applied to transmission systems, such as the estimator proposed in Baran and Kelley (1994). However, some characteristics of DSs make it very difficult to apply this type of estimator in such systems. According to Baran (2012) and Lefebvre et al. (2014), these characteristics are as follows:

- DSs are usually composed of radial feeders, with branches that can be single-phase, two-phase (residential loads), or three-phase (commercial and industrial loads).
- DSs are composed of nontransposed power lines, naturally unbalanced (due to the impossibility of evenly distributing loads in the feeders) and with high resistance/reactance ratio.
- There are several possibilities of connection of the transformers and the loads of the DSs.
- The size of the problem is very high, since even for a small city (up to 50,000 inhabitants), DS may contain thousands of two-phase and three-phase buses.
- Most feeders have current magnitude measurements instead of power measurements, which represents a problem for the use of conventional estimators (numerical problems).
- Usually the status of the capacitor banks and disconnect device, as well as the transformer tap position, are not usually monitored directly.
- Reduced number of real-time measurements in comparison to transmission systems.
- Use of measurements with very different accuracies and sampling rates depending on the different measurement equipment available, such as PMUs and SCADA systems. It is also important to mention the virtual measurements, from zero-injection buses, and pseudo measurements of loads from advanced metering infrastructure or historical system operation data (Ghosh et al., 1997; Arefi et al., 2015).

These additional complexities have fostered the development of specialized estimators for DSs, which use formulations that are designed to suit these characteristics (Baran and Kelley, 1994; Deng et al., 2002; Baran and McDermott, 2009; de Almeida and Ochoa, 2017).

The following are three types of estimators, developed for DSs, which stand out in the literature.

4.7.1 State estimator based on power flow calculation

In the approach presented in Deng et al. (2002), the calculation of the power flow for electrical networks is done by integrating the measurements available in real-time during the iterations of the power flow. In the case of DSs, the forward/backward sweep method is used to calculate the power flow associated with the solution of the normal equations in the branches that have measurements. In addition, the values of the pseudo measurements, initially generated through a demand allocation algorithm, are adjusted to obtain power flow results close to the monitored values.

The general steps of the algorithm of this estimator follow below:

1. Initialization of the voltages and demands: An initial voltage value is assumed for all the buses (flat start or measured values if available), and the initial values of the power injection pseudo measurements obtained through an algorithm of demand allocation.
2. Backward sweep: Power flows are calculated for all branches from the leaf nodes (end buses) to the root node (substation) by the sum of the power injections and the loss calculation in the branches. In monitored branches, the measurement model of Eq. (4.2) is used, making an estimation only for this branch, considering as reference the upstream bus. With the state estimated by Eq. (4.15) (normal equation), the equivalent power flow of the monitored branch is calculated.
3. Forward sweep: The nodal voltages are updated from the root node (substation) to the leaf nodes (end buses) using the Ohm's Law and the flows calculated in the Backward sweep.
4. Adjustment of the pseudo measurements: Calculates and applies adjustment factors in the pseudo measurements of power injection obtained based on the differences between the measured and estimated values of these flows.
5. Convergence test: If the power difference in the buses between two consecutive iterations is less than the preestablished tolerance, return the estimated state. Otherwise, return to Step 2.

4.7.2 State estimator considering branch current flows as state variables

First presented in Baran and Kelley (1994), the formulation of this estimator uses the WLS method to estimate the state variables of the network. However, instead of the traditional formulation, which uses complex nodal voltages as state variables, in this case, the complex current flows in the branches of the distribution network are used as state variables in rectangular or polar coordinates. According to the authors, among the advantages of this approach, the estimation process can be decoupled between the phases and does not present problems of numerical conditioning (Baran and McDermott, 2009).

The measurement model is then constructed according to Eq. (4.2) but considering as state vector the complex current flows between buses expressed in rectangular coordinates. The measurement model, presented in Eq. (4.2), is then constructed using equivalent current measurements, according to each type of measurement, relating them to the current state variables. That is, flow and power injection, voltage magnitude, or synchronized phasor measurements are converted into equivalent current measurements (Wang and Schulz, 2004). With this, the solution of the normal Eq. (4.15) is used to update the state variables.

The general steps of the algorithm of this estimator follow below:

1. By performing a backward sweep, the values of the state variables (current flows in the branches) are initialized and the nodal voltages are initialized by a forward sweep.
2. Obtain and update equivalent current measurements.
3. Obtain the correction in the current state variables through the normal Eq. (4.15).
4. With the new currents obtained, update the nodal voltages through the forward sweep method for radial networks.
5. Convergence test: If the difference between the current flows of two consecutive iterations is less than the preestablished tolerance, return the estimated state. Otherwise, return to Step 2.

4.7.3 State estimator based on admittance matrix

The method also uses the measurement model of Eq. (4.2) but using only equivalent measurements of current and complex voltage expressed in rectangular coordinates (Lu et al., 1995). Like the previous method, the flow measurements and power injections are transformed into equivalent current measurements. In this method, the Jacobian matrix remains constant and is composed of the elements of the nodal admittance matrix of the electric system. The estimated state, composed of the nodal complex voltages, is also obtained through the solution of the normal Eq. (4.15).

The general steps of the algorithm of this estimator follow:

1. An initial voltage value is assumed for all the buses (flat start or measured values, if available) and the available values of the power injection pseudo measurements.
2. Convert the flow and injection measurements of power and magnitude of voltage into equivalent current measurements.
3. Calculate the Jacobian and Gain matrices that will be kept constant in the estimation process.
4. Obtain the correction in the state variables of nodal voltages in the rectangular form through the normal Eq. (4.15).
5. Convergence test: If the difference between the state vector of two consecutive iterations is less than the tolerance returns the estimated state. Otherwise, return to Step 4.

4.7.4 Crucial challenges for DS state estimation

It is worth noting that, although specialized estimators for DSs have already been developed, a consensus on the best method for estimating DSs status has not yet been reached, partly because of the few reports of practical application in the literature (Svenda et al., 2017; Massignan et al., 2018).

In a general way, due to the particularities of DS networks, the proposed specialized estimators for DSs not only involve approximation in the solution process but also result in approximation to the original problem (Feng et al., 2012). The simplifications and approximations of the measurement model usually assumed by those specialized estimators make difficult to generalize their results. Within this context, an efficient and numerically stable algorithm to perform three-phase state estimation in DSs using the traditional WLS nonlinear formulation is presented in Hebling et al. (2020). It uses nodal voltages as the state variables and does not require any additional assumption from the traditional state estimation formulation. Its numerical stability is achieved through the application of QR-based formulation in the WLS algorithm, while computational efficiency is obtained through sparse routines based on the Multifrontal approach. The algorithm has been successfully applied to IEEE three-phase unbalanced test feeders (IEEE34, IEEE123, and IEEE US low voltage (342 nodes)).

In terms of the state of the art of the state estimation problem in DSs, the following points stand out as crucial challenges according to Gomez-Expósito et al. (2011); Lefebvre et al. (2014); Primadianto and Lu (2017); and Ahmad et al. (2018):

- Distribution automation and smart meters: new equipment and communication standards, synchronization challenges between different types of measurements.
- Modeling of pseudo measurements: statistical models for pseudo measurements, incorporation of smart meters, consideration of distributed generation in pseudo measurement modeling, correlations between pseudo measurements, modeling of the low-voltage system.
- Wide-area monitoring: multiarea state estimation, efficient numerical methods, computational performance for real-time applications.
- Increase in the insertion of distributed generation in both the medium (usually wind) and low voltage (usually solar photovoltaic).
- Treatment of topology errors caused by incorrect information on the status of switches and circuit breakers.

4.8 Exercises

1. In the direct-current circuit shown in Fig. 4.4, all the currents in the branches and all the nodal voltages (currents in Ampere and voltages in Volts) are measured.

Considering the measurement vector being given by

$$z = [I:1-2\ \ I:2-1\ \ I:1-3\ \ I:3-1\ \ I:2-3\ \ I:3-2\ \ V:1\ \ V:2\ \ V:3]^T$$

$$z = [52\ \ -49\ \ -15\ \ 16\ \ -82\ \ 80\ \ 11\ \ 6\ \ 14]^T$$

(a) Determine the state vector (nodal magnitude voltages) by the WLS estimator (considering R equal to an identity matrix).

(b) Assuming that the errors of measurements *I: 1–2* and *I: 1–3* have variances 100 times smaller than the others, repeat item (a) using the WLS estimator.

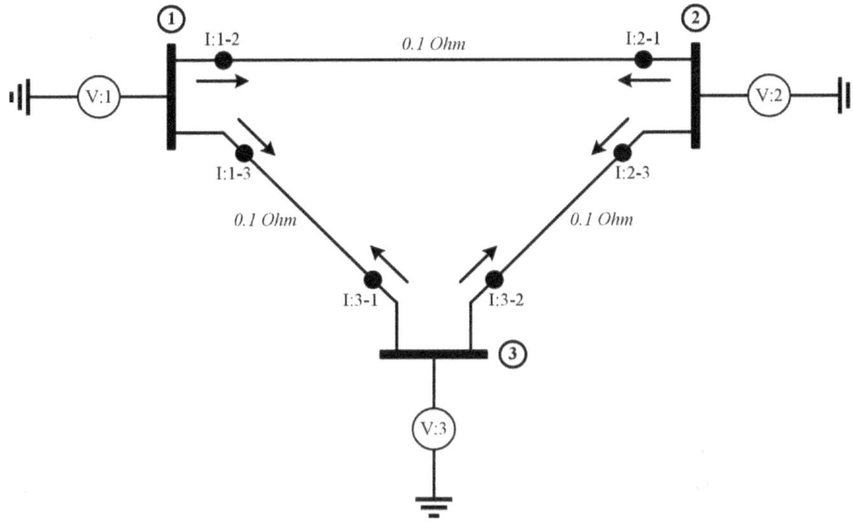

Fig. 4.4 DC three-bus power system.

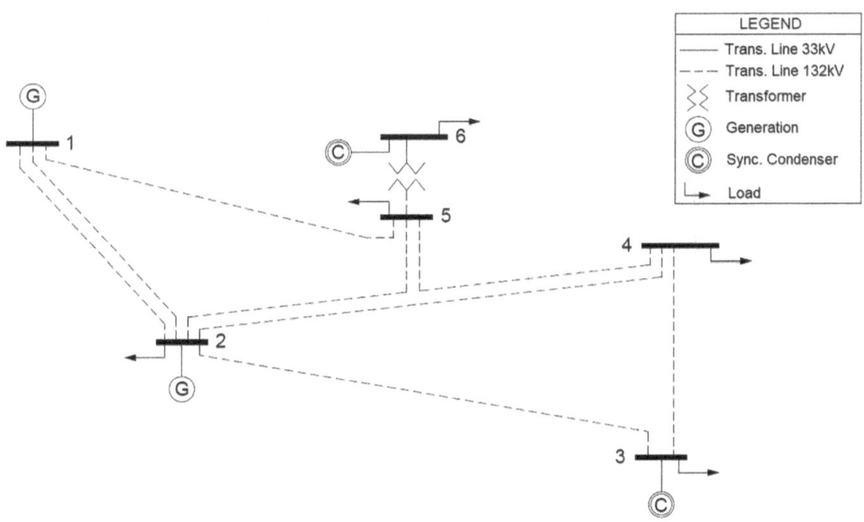

Fig. 4.5 Six-bus power system.

2. Consider the six-bus electrical system shown in Fig. 4.5. The available measurements and corresponding values are presented in Tables 4.4 and 4.5. The parameters of this system are shown in Table 4.6. With these data, estimate the state variables of this system through the WLS estimator.

Table 4.4 Injection measurements of power and voltage in buses.

Bus	P_k (pu)	Q_k (pu)	V_k (pu)
1	1.4902	−0.0368	1.0600
2	0.1830	0.11199	1.0450
3	−0.9421	0.0221	1.0100
4	−0.4779	0.0390	1.0250
5	−0.0760	−0.0158	1.0280
6	−0.11120	−0.1399	1.0700

Table 4.5 Branch power flow measurements.

From (k)	To (l)	P_{kl} (pu)	Q_{kl} (pu)	P_{lk} (pu)	Q_{lk} (pu)
1	2	1.0310	−0.0694	−1.0127	0.0669
1	5	0.4592	0.0325	−0.4489	−0.0436
2	3	0.6248	0.0474	−0.6078	−0.0220
2	4	0.3428	−0.0057	−0.3366	−0.0118
2	5	0.2281	0.0112	−0.2253	−0.0400
3	4	−0.3343	0.0440	0.3418	−0.0381
4	5	−0.4831	0.0889	0.4862	−0.0793
5	6	0.1120	0.1470	−0.1120	−0.1399

Table 4.6 Electrical system parameters.

From (k)	To (l)	r (pu)	x (pu)	b^{sh} (pu)	tap (pu)
1	2	0.0194	0.0592	0.0264	1
1	5	0.0540	0.2230	0.0246	1
2	3	0.0470	0.1980	0.0219	1
2	4	0.0581	0.1763	0.0170	1
2	5	0.0570	0.1379	0.0173	1
3	4	0.0670	0.1710	0.0064	1
4	5	0.0134	0.0421	0.0000	1
5	6	0.0000	0.2520	0.0000	1.0730

Note: Consider that all available measurements have standard deviations given by

$$\sigma_i = |z_i|/100,$$

where z_i is the value of measurement i.

References

Abur, A., Expósito, A., 2004. Power Systems State Estimation: theory and implementation. Marcel Dekker, Inc., New York, USA.
Ahmad, F., Rasool, A., Ozsoy, E., Sekar, R., Sabanovic, A., Elitas, M., 2018. Distribution system state estimation—a step towards smart grid. Renew. Sustain. Energy Rev. 81, 2659–2671.
Almeida, M., Asada, E., Garcia, A., 2006. Effects of Load Imbalance and System Asymmetry on Three-Phase State Estimation. IEEE PES General Meeting, Montreal, Canada, pp. 1–6.
Arefi, A., Ledwich, G., Behi, B., 2015. An efficient DSE using conditional multivariate complex Gaussian distribution. IEEE Trans. Smart Grid 6 (4), 2147–2156.
Baran, M., 2012. Branch Current Based State Estimation for Distribution System Monitoring. IEEE PES General Meeting, San Diego, USA, pp. 1–4.
Baran, M.E., Kelley, A.W., 1994. State estimation for real-time monitoring of distribution systems. IEEE Trans. Power Syst. 9 (3), 1601–1609.
Baran, M., McDermott, T.E., 2009. State Estimation for Real Time Monitoring of Distribution Feeders. Canada, IEEE PES General Meeting, Calgary. p.1-4.
Costa, A.S., Quintana, V., 1981a. A robust numerical technique for power system state estimation. IEEE Trans. Power App. Syst. 100 (2), 691–698.
Costa, A.S., Quintana, V., 1981b. An orthogonal row processing algorithm for power system sequential state estimation. IEEE Trans. Power App. Syst. 100 (8), 3791–3800.
Costa, A.S., Albuquerque, A., Bez, D., 2013. An estimation fusion method for including phasor measurements into power system real-time modeling. IEEE Trans. Power Syst. 28 (2), 1910–1920.
de Almeida, M.C., Ochoa, L.F., 2017. An improved three-phase AMB distribution system state estimator. IEEE Trans. Power Syst. 32 (2), 1463–1473.
Deng, Y., He, Y., Zhang, B., 2002. A branch-estimation-based state estimation method for radial distribution systems. IEEE Trans. Power Deliv. 17 (4), 1057–1062.
Ebrahimian, R., Baldick, R., 2001. State estimator condition number analysis. IEEE Trans. Power Syst. 16 (2), 273–279.
Feng, X., Yang, F., Peterson, W., 2012. A practical multi-phase distribution state estimation solution incorporating smart meter and sensor data. In: IEEE Power and Energy Society General Meeting.
Ghosh, A.K., Lubkeman, D.L., Downey, M.J., Jones, R.H., 1997. Distribution circuit state estimation using a probabilistic approach. IEEE Trans. Power Syst. 12 (1), 45–51.
Gjelsvik, A., Aam, S., Holten, L., 1985. Hachtel's augmented matrix method—a rapid method improving numerical stability in power system static state estimation. IEEE Trans. Power App. Syst. 104 (11), 2987–2993.
Golub, G., van Loan, C., 1996. Matrix Computations, third ed. The John Hopkins University Press, Baltimore, USA.
Gomez-Expósito, A., Abur, A., de la Villa Jaen, A., Gomez-Quiles, C., 2011. A multilevel state estimation paradigm for smart grids. Proc. IEEE 99 (6), 952–976.
Hansen, C., Debs, A., 1995. Power system state estimation using three-phase models. IEEE Trans. Power Syst. 10 (2), 818–824.
Hebling, G.M., Massignan, J.A.D., London Jr., J.B.A., Camillo, M.H.M., 2020. Sparse and numerically stable implementation of a distribution system state estimation based on multifrontal QR factorization. Electr. Pow. Syst. Res. 189, 106734.
Holten, L., Gjelsvik, A., Aam, S., Wu, F., Liu, W., 1988. Comparison of different methods for state estimation. IEEE Trans. Power Syst. 3 (4), 1798–1806.
Horisberger, H., Richard, J., Rossier, C., 1976. A fast-decoupled static state estimator for electric power systems. IEEE Trans. Power App. Syst. 95 (1), 208–215.

Irving, M., Owen, R., Sterling, M., 1978. Power-system state estimation using linear programming. Proc. Inst. Electr. Eng. 125 (9), 879–885.
Korres, G.N., Manousakis, N.M., 2011. State estimation and bad data processing for systems including PMU and SCADA measurements. Electr. Power Syst. Res. 81 (7), 1514–1524.
Lefebvre, S., Prévost, J., Lenoir, L., 2014. Distribution State Estimation: A Necessary Requirement for the Smart Grid. IEEE PES General Meeting, National Harbor, USA, pp. 1–5.
London Jr., J.B.A., Piereti, S.A.R., Benedito, R.A.S., Bretas, N.G., 2009. Redundancy and observability analysis of conventional and PMU measurements. IEEE Trans Power Syst 24, 1629–1630.
Lu, C.N., Teng, J.H., Liu, W.E., 1995. Distribution system state estimation. IEEE Trans. Power Syst. 10 (1), 229–240.
Manitsas, E., Singh, R., Pal, B.C., Strbac, G., 2012. Distribution system state estimation using an artificial neural network approach for pseudo measurement modeling. IEEE Trans. Power Syst. 27 (4), 1888–1896.
Massignan, J.A.D., London Jr., J., Bessani, M., Maciel, C.D., Delbem, A.C.B., Camillo, M.-H.M., Lima Soares, T.W., 2018. In-field validation of a real-time monitoring tool for distribution feeders. IEEE Trans. Power Deliv. 33 (4), 1798–1808.
Meliopoulos, A., Cokkinides, G., Stefopoulos, G., 2005. Numerical Experiments for Three-Phase State Estimation Performance and Evaluation. IEEE Powertech Conference, St. Petersburg, Russia, pp. 1–7.
Mili, L., Phaniraj, V., Rousseeuw, P., 1991. Least median of squares estimation in power systems. IEEE Trans. Power Syst. 6 (2), 511–523.
Mili, L., Cheniae, M., Rousseeuw, P., 1994. Robust state estimation of electric power systems. IEEE Trans. Circ. Syst. I: Fund. Theory Appl. 41 (5), 349–358.
Monticelli, A., 1999. State Estimation in Electric Power Systems: A Generalized Approach. Kluwer Academic Publishers, Boston, USA.
Monticelli, A., 2000. Electric power system state estimation. Proc. IEEE 88 (2), 262–282.
Monticelli, A., Garcia, A., 1990. Fast decoupled state estimators. IEEE Trans. Power Syst. 5 (2), 556–564.
Nucera, R., Gilles, M., 1991. Observability analysis: a new topological algorithm. IEEE Trans. Power Syst. 6 (2), 466–475.
Primadianto, A., Lu, C., 2017. A review on distribution system state estimation. IEEE Trans. Power Syst. 32 (5), 3875–3883.
Quarteroni, A., Sacco, R., Saleri, F., 2000. Numerical Mathematics, first ed. Springer, New York, USA.
Schweppe, F., Handschin, E., 1974. Static state estimation in electric power systems. Proc. IEEE 62 (7), 972–982.
Svenda, G., Strezoski, V., Kanjuh, S., 2017. Real-life distribution state estimation integrated in the distribution management system. Int. Trans. Electr. Energy Syst. 27 (5), 1–16.
Wang, H., Schulz, N.N., 2004. A revised branch current-based distribution system state estimation algorithm and meter placement impact. IEEE Trans. Power Syst. 19 (1), 207–213.
Wu, F., Monticelli, A., 1985. Network observability: theory. IEEE Trans. Power App. Syst. 104 (5), 1042–1048.
Zhao, J., Mili, L., 2018. Vulnerability of the largest normalized residual statistical test to leverage points. IEEE Trans. Power Syst. 33 (4), 4643–4646.
Zhong, A., Abur, A., 2002. Effects of nontransposed lines and unbalanced loads on state estimation. In: IEEE Power Engineering Society Winter Meeting, 2002 New York, USA. vol. 2, pp. 975–979.

CHAPTER 5

Qualitative characteristics of measurement sets

In the context of static state estimation in EPS, qualitative characteristics of measurement sets refer to the analyses of observability and redundancy of measurements. In this way, this chapter will present concepts, definitions, and some methods already developed to treat problems related to observability and redundancy of measurements.

The qualitative characteristics of measurement sets are important information both in the design phase of metering systems and in real-time operation. For the design of metering systems, they should be considered to obtain observable metering systems with adequate level of redundancy to enable noise filtering and GE processing. In terms of real-time operation, the qualitative characteristics of sets of measurements should be updated as the analog measurements provided to the state estimator may be temporarily inaccessible to the estimator. That might occur due to some problem in the telemetry system causing the unavailability of communication channels.

In this chapter, the analysis of observability and redundancy of measurements will be approached in detail considering the single-phase modeling of the power grid, the existence of only SCADA measurements and the state variables to be estimated are complex nodal voltages. This is because most of the estimators and methods developed for observability and redundancy analysis work under these considerations. It is worth remembering, however, that the methods that will be presented in this chapter for analysis of observability and redundancy of measurements can be extended to consider the three-phase modeling of the network as well as the existence of SCADA measurements and SPM, as it will be discussed in the last sections of this chapter.

5.1 Observability analysis

5.1.1 Introduction

As mentioned in Chapter 1, the problem of observability consists of verifying that it is possible, through the measurements available, to estimate all the

Cyber-Physical Power Systems State Estimation
https://doi.org/10.1016/B978-0-323-90033-1.00001-9

state variables of an EPS. In the positive case, the EPS is said to be observable and state estimation is performed for the whole system. Otherwise, the EPS is not observable as a whole, and two alternatives arise to solve the problem:

- Identify the observable islands of the EPS and estimate their state variables separately. An observable island is defined as a set of EPS buses between which the values of the flows can be determined from the available measurements.
- Restore observability by selecting critical pseudo measurements. The term pseudo measurement has already been defined in Chapter 1 and it refers to historical data, load forecasting and generation data, etc., which are available at the operating centers. A pseudo measurement is critical if it is the only measurement providing the information of a state variable and, therefore, its measured and estimated values are equal (zero residual). It aims to restore observability through critical pseudo measurements so that they do not affect the estimates of measurements obtained in real time, which are usually more accurate.

In Section 5.1.2, important definitions will be presented for understanding the other sections of this chapter.

5.1.2 Definitions

In Clements and Wollember (1975), the minimum conditions for observability of an EPS were investigated, using its topology and the Kirchhoff laws. The concept of observable islands was then defined and a heuristic algorithm was proposed that served as the basis for the elaboration of the concepts of algebraic, topological, and numerical observability presented in Krumpholz et al. (1980) and defined as a function of the formulation of the WLS state estimator. These concepts will be presented in the following, as well as the concept of observability $P\theta$ and QV.

Algebraic observability

By analyzing the formulation of the WLS state estimator presented in Chapter 4, it is verified that, in order to solve the Normal Eq. (4.15) and to obtain a single vector $\hat{x}$, the Gain matrix, given by Eq. (4.16), must be nonsingular, that is, its rank must be equal to the number of state variables to be estimated.

Since the matrix R^{-1} is a nonsingular diagonal matrix, it must be noted that for a system to be observable, the Jacobian matrix H must have rank equal to the number of state variables to be estimated. In this case, the system will be "algebraically" observable.

Once the concept of algebraic observability has been defined, the application of the $P\theta$ (active) and QV (reactive) decoupling to the problem of observability allows its decomposition in a similar way, as will be shown next.

Pθ *and* QV *observability*

Making use of the decoupling model, also known as $P\theta$ and QV decoupling, the fast decoupled power flow method proposed in Sttot and Alsac (1974), as well as the fast decoupled state estimation method presented in Garcia et al. (1979), can obtain computational advantages in the solution of their respective problems. The $P\theta$ and QV decoupling can also be used to facilitate the analysis of the observability of EPS.

For an EPS with n buses, the number of state variables to be estimated is $2n-1$, with $n-1$ voltage phase angles and n bus voltage magnitudes, considering that one of the buses of the system is adopted as the angular reference. Thus to determine if an EPS with n buses is observable, it is important to verify if it is possible to estimate all $2n-1$ state variables considering all available measurements. Consequently, by making use of the previously presented definition of algebraic observability, an EPS with n buses is algebraically observable with respect to a set of measurements if.

$$Rank\,(H) = 2n - 1. \tag{5.1}$$

However, through the $P\theta$ and QV decoupling, which is based on the fact that the sensitivities $\frac{\partial P}{\partial \theta}$ and $\frac{\partial Q}{\partial V}$ are more intense than the sensitivities $\frac{\partial P}{\partial V}$ and $\frac{\partial Q}{\partial \theta}$, it is possible to perform the analysis of observability separately for each of the models.

Thus in order to determine whether an EPS is $P\theta$ observable, it is necessary to verify if it is possible to estimate its $n-1$ voltage phase angles considering only the active power measurements (power flows and injections). Similarly, in order to determine if the EPS is QV observable, it is necessary to verify if it is possible to estimate the n voltage magnitudes of this EPS considering only the reactive power measurements (power flows and injections) and voltage magnitude measurements in buses.

Considering the definition of algebraic observability, an EPS with n buses is algebraically $P\theta$ observable given a set of measurements if.

$$Rank\left(H_{p\theta}\right) = n - 1. \tag{5.2}$$

Likewise, an EPS with n buses is algebraically QV-observable with respect to a set of measurements if.

$$Rank\left(H_{qV}\right) = n. \tag{5.3}$$

It is worth mentioning that for metering systems in which the active and reactive power measurements are performed in pairs, the $P\theta$ algebraic observability guarantees the QV algebraic observability, if there is at least one measurement of voltage magnitude. Thus for metering systems with these characteristics, observability analysis can be performed considering only the $P\theta$ coupling (Krumpholz et al., 1980; Wu and Monticelli, 1985; Monticelli and Wu, 1985).

Numerical observability

An EPS is said "numerically" observable with respect to a set of measurements if it is possible to estimate the vector of state variables through the iterative solution of the normal Eq. (4.15), starting from the flat start.

Note 5.1: For the normal equation to have a solution, it is necessary that the rank of the Jacobian matrix H is equal to the number of state variables to be estimated, that is, the system must be algebraically observable. However, for the system to be numerically observable, the matrix H must be numerically well conditioned, that is, far enough from its region of singularity. This last characteristic defines the concept of numerical observability. Consequently, although numerical observability ensures algebraic observability, the reciprocal is not always true.

Topological observability

The concept of topological observability explores the $P\theta$ and QV decoupling and is based on the relationship between measurements and the network graph, whose edges and vertices represent the branches and the buses of the system, respectively. According to this relation, an EPS is "topologically" observable, given a set of measurements, only if there exists, associated with such a system, a full-rank spanning tree (spanning tree is a tree covering all the buses of a system, a spanning tree is said full-rank if at least one distinct measurement can be assigned to each branch). A full-rank spanning tree can be simply called an observable spanning tree (Krumpholz et al., 1980).

In order to verify the topological observability, it is necessary to obtain an observable spanning tree. To do so, a search process is performed, which walks through the network topology (network graph) associating, for each branch, a different measurement. The following rule was used for the association of measurements for network graph edges (Krumpholz et al., 1980; Quintana et al., 1982): a measurement z can be assigned to an edge a of the graph if z is a flow measurement in the branch corresponding to that edge or an injection measurement in one of the end buses of that branch. In order to consider the measurement of voltage magnitude in the analysis of topological observability, it has been demonstrated that, because this measurement gives information of the magnitude of voltage in the bus where it was carried out, it can be treated as a measurement of reactive power flow in a fictitious unit susceptance branch, connecting the bus where the voltage measurement occurs, to an additional bus representing the ground node (voltage reference). Thus topological observability analysis is performed considering only two types of measurements, flow and injection measurements.

Therefore $P\theta$ and QV topological observability analyses are performed separately but treated in a similar way (Van Cutsem, 1985). However, for metering systems with parity between active and reactive power measurements with at least one measurement of magnitude of voltage, the topological observability analysis can be performed only once, considering only the $P\theta$ model.

According to the reference literature (Monticelli, 1999; Abur and Expósito, 2004), in general, topological methods for observability analysis are subject to a combinatorial explosion, considering that they require search procedures of an observable spanning tree. The computational cost to perform this search becomes high due to the existence of injection measurements, considering that this type of measurement relates the complex voltage of the bus in which it was performed with the complex voltages of the other buses that are adjacent. To exemplify what was said, we will analyze the $P\theta$ topological observability of the power system presented in Fig. 5.1. Note that the active power measurements, shown in that figure, guarantee the $P\theta$ observability of the system.

After applying the search procedure to the system of Fig. 5.1, it is verified that the injection measurement in bus 2 can be attributed to the branches connecting the buses: 2 and 1; 2 and 3; and 2 and 5. If that measurement is assigned to the branch connecting nodes 2 and 3, the search process would not obtain an observable spanning tree for that system, because measurement $P{:}1$ (power injection in bus 1) would be attributed to the branch between buses 1 and 2; measurement P:2 would be attributed to the branch between the buses 2 and 3; measurement $P{:}3-4$ (power flow from bus 3 to bus 4) would be attributed to the branch between buses 3 and 4; and measurement

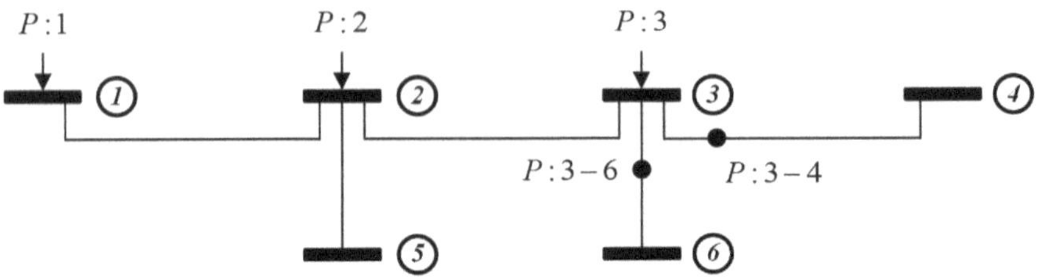

Fig. 5.1 Example system of six buses associated with a set of five active power measurements (•, flow measurements; →, injection measurements).

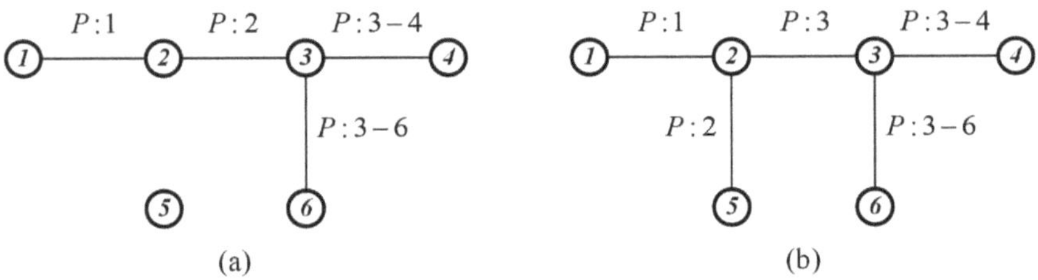

Fig. 5.2 Construction of an observable spanning tree for the example system of six buses.

$P{:}3-6$ would be attributed to the branch between buses 3 and 6. Note that the result of this search would not be an observable spanning tree, since bus 5 was isolated, as shown in Fig. 5.2A. In this situation, if the possibility of associating the measurement $P{:}2$ with another branch of the network was verified, the search process would return to bus 2 and assign the measurement $P{:}2$ to the branch between buses 2 and 5. Then the search process would find an observable spanning tree, as shown in Fig. 5.2B. Consequently, whenever there are power injection measurements, the possibility of paths to be analyzed in the search for an observable spanning tree will be increased (it is worth mentioning that topological methods make use of some heuristic techniques to improve this search process).

Another disadvantage of topological methods, highlighted in the literature by advocates of the so-called numerical methods for observability analysis, is that topological methods require specific routines that are not available in computational applications for state estimation. On the other hand, proponents of topological methods point out that these methods are not subject to numerical errors, since they do not require numerical calculations.

The authors of this book believe, however, that topological analysis is the best way to understand the problem of observability in the context of PSSE because, by thinking in terms of an observable spanning tree, it becomes possible to "see" the observability problem, walking in the topology of the network through the association of branches and measurements.

Note 5.2: In Krumpholz et al. (1980), the equivalence of algebraic and topological observability has been demonstrated for several practical situations of interest.

Observable and unobservable branches

Observable (unobservable) branches are those whose power flows can (cannot) be determined from the available measurements.

Irrelevant injection measurements

Irrelevant injection measurements for the observability of EPS are measurements incident to at least one unobservable branch. These measurements should not be considered for the estimation of the state of the observable islands. That is why they are called discardable measurements for state estimation of observable islands.

5.1.3 Methods

The methods developed for the treatment of the observability problem can be grouped, in general, into topological and numerical ones.

According to what was presented in the previous section, topological methods (Krumpholz et al., 1980; Quintana et al., 1982; Mori and Tsuzuki, 1991) are based upon concepts of graph theory and they require the creation of specific routines, which do not require calculations but are of complex and combinatorial nature (Monticelli, 1999; Abur and Expósito, 2004). In turn, numerical methods are generally easier to implement in a computer (Wu and Monticelli, 1985; Monticelli and Wu, 1985; Bretas, 1996; Castillo et al., 2006; Almeida et al., 2008), since most of them aim at the use of routines already available in state estimation software. However, they are subject to numerical errors.

Among the numerical methods, the method proposed in Monticelli and Wu (1985), which is based on the triangular factorization of the Gain matrix, stands out as one of the most cited. In case the system is not completely observable, to proceed to the identification of observable islands or to the restoration of observability (via pseudo measurements), this method requires the solution of a system of algebraic equations for the determination of the unobservable branches.

In an attempt to reduce the amount of calculations required, a numerical method for observability analysis was proposed in Bretas (1996), which is also based on the triangular factorization of the Gain matrix. However, making use of the concept of factorization paths (Tinney et al., 1985), such a method does not require the solution of algebraic equation systems for the identification of observable islands or for restoration of observability.

Because the Jacobian matrix H is numerically better conditioned than the Gain matrix, in Monticelli and Wu (1986) a method for observability analysis has been developed through the triangular factorization of matrix H. Since it is an extension of the methodology proposed in Wu and Monticelli (1985), this method also requires the solution of systems of algebraic equations.

A method that allows observability analysis and identification of Critical Measurements (CMs), through the row-echelon reduction of the Jacobian matrix, is found in Falcao and Arias (1994), which, however, does not allow the identification of observable islands.

The numerical method for observability analysis presented in Almeida et al. (2008) is based on the factorization of the so-called Gram matrix (Pussell and Trimble, 1991), which is obtained from the matrix product of matrix H by its transpose. To identify observable islands, such a method requires the solution of a system of algebraic equations by obtaining and factorizing the Gram matrix.

In general, the idea behind the development of numerical techniques for observability analysis is to obtain computer-friendly procedures that require a small number of calculations. In this context, we highlight the method developed in Benedito et al. (2009), which is also based on a triangular factorization process. However, instead of performing triangular factorization of the Gain matrix or the Gram matrix, both obtained from matrix products involving the H matrix, the method performs the triangular factorization directly from the matrix H.

Next, two methods will be presented for the treatment of the observability problem, both of which do not require the solution of algebraic equation systems and are based on the triangular factorization process and the theory of factorization paths.

Method based on the triangular factorization of the Gain matrix and the concept of factorization paths

The method proposed by Bretas (1996), for observability analysis, is based on the triangular factorization of the WLS Gain matrix and the concept of factorization paths (Tinney et al., 1985). If the system is not observable as a whole, to identify the observable islands or restore observability, the method makes use of the factorization paths associated with the triangular factorization of that matrix. In this way, such tasks can be performed without requiring any solution of algebraic equation systems.

Here are some important definitions and properties to understand how the method works. For this, the Gain matrix of the linear WLS state estimator presented in Chapter 4 will be considered. In this sense, it is important to remember that the optimal solution for this estimator can be calculated by

$$\hat{\theta} = G_{\theta}^{-1} H_{DC}^{T} R_{ap}^{-1} z_{ap}. \tag{5.4}$$

where the matrix G_{θ} is defined by the following equation:

$$G_{\theta} = H_{DC}^{T} R_{ap}^{-1} H_{DC} \tag{5.5}$$

The other terms that appear in these equations were defined in Chapter 4.

Note 5.3: It is worth noticing that performing analysis of observability or redundancy of measurements through the Gain matrix of the linear WLS state estimator is equivalent to the analysis of the observability of the $P\theta$ model. This is because the Gain matrix of this model is equivalent to the G_{θ} matrix. Similarly, the matrix $H_{P\theta}$ is equivalent to the H_{DC} matrix.

Triangular factorization of the Gain matrix

In Wu and Monticelli (1985), the authors demonstrated that the observability of an EPS can be investigated through the relationship between reference angles in the system and the triangular factorization of the G_{θ} matrix. This is because for a system with n buses to be observable the matrix G_{θ} must have rank equal to $n-1$.

Considering the above, the matrix resulting from the triangular factorization of the matrix G_{θ} associated with an observable EPS will only have one null pivot, at the last diagonal. This indicates the need for an angular reference, as presented in the structure shown in the following figure (the dark area corresponds to possible nonzero elements) (Fig. 5.3):

If the EPS is not completely observable, the corresponding matrix G_θ will have rank lower than $n-1$. Thus during the factorization of this matrix, a null pivot will appear on the ith diagonal, where $i<n$, indicating that the system is not observable as a whole, and the other elements of row and column i will be null (as shown in Wu and Monticelli, 1985). The matrix resulting from this factorization will have the following form (Fig. 5.4):

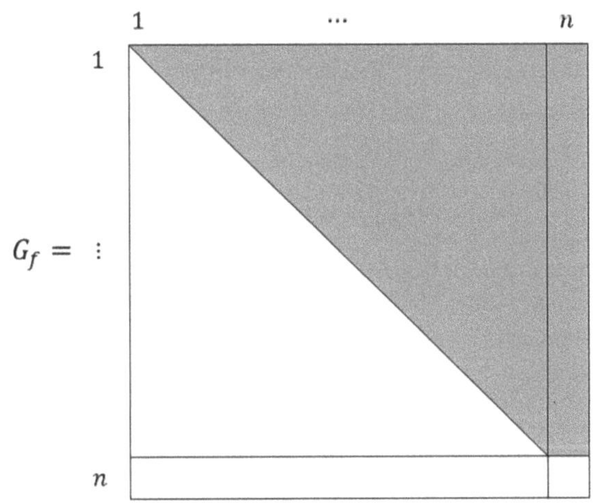

Fig. 5.3 Factored Gain matrix of an observable EPS with n buses.

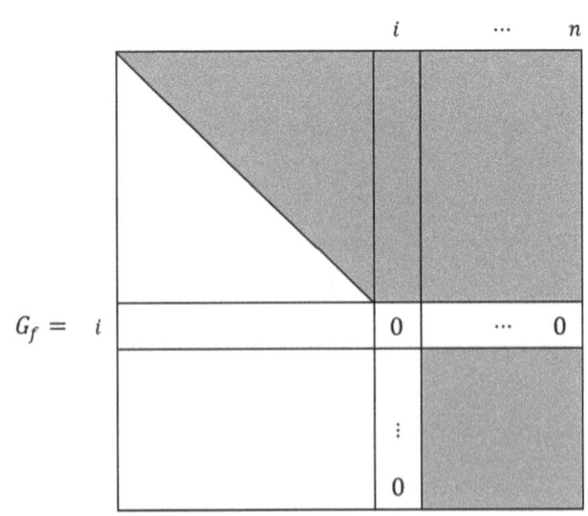

Fig. 5.4 Triangular factorization of the Gain matrix of an unobservable system.

Factorization paths

The concept of factorization paths was presented in Tinney et al. (1985), with the main objective of improving the solution efficiency of systems of algebraic equations of type $Ax=b$, in which, in addition to the matrix of the coefficients A being sparse, the independent vector b is sparse and/or it is intended to determine only some positions of the dependent vector x. This concept was applied for the first time to the problem of analysis of the observability of EPS in Bretas (1996), taking advantage of the fact that the factorization paths indicate the relations of linear dependence between the rows of the factored matrix.

The factorization paths are obtained from the Lu factorization of a matrix, that is, from the triangular factors responsible for matrix factorization. For example, consider the matrix G_θ corresponding to the six-bus EPS associated with a measurement set as shown in Fig. 5.5. Note that for this measurement set the EPS is observable.

To facilitate obtaining of the matrices, the reactances of all six-bus EPS branches shall be considered as equal to 1 pu and the weight matrix is an identity matrix. It is worth mentioning that these considerations are usually assumed by the methods based on the triangular factorization of the Gain matrix. This is because the qualitative characteristics of measurement sets do not depend on network parameters and measurements weights, but only on the type and location of measurements. The G_θ matrix of this system is as follows:

$$G_\theta = \begin{bmatrix} 5 & -3 & -2 & 0 & 0 & 0 \\ -3 & 2 & 1 & 0 & 0 & 0 \\ -2 & 1 & 3 & -4 & 1 & 1 \\ 0 & 0 & -4 & 11 & -4 & -3 \\ 0 & 0 & 1 & -4 & 2 & 1 \\ 0 & 0 & 1 & -3 & 1 & 1 \end{bmatrix} \tag{5.6}$$

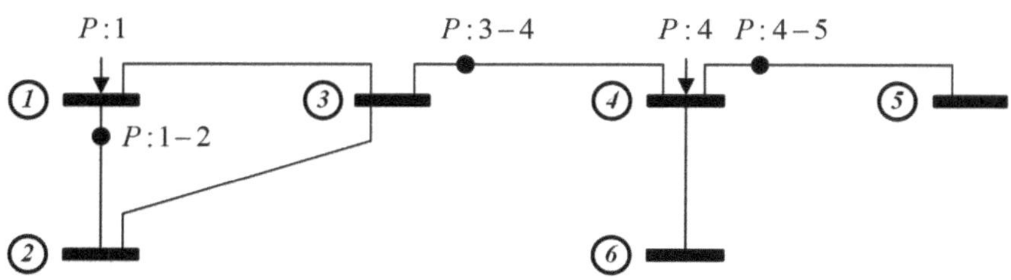

Fig. 5.5 Six-bus system associated with an observable measurement set (•, active flow measurements; →, active injection measurements).

With the LU decomposition of this matrix, we obtain the matrix of the triangular factors U and plot the factorization paths from their nonzero elements, as can be visualized through the arrows indicated in matrix U shown below:

$$U_{G\theta} = \begin{bmatrix} 1 & \rightarrow & -0.6 & & -0.4 & & 0 & & 0 & & 0 \\ & & \downarrow & & & & & & & & \\ 0 & & 1 & \rightarrow & -0.999 & & 0 & & 0 & & 0 \\ & & & & \downarrow & & & & & & \\ 0 & & 0 & & 1 & \rightarrow & -2 & & 0.5 & & 0.5 \\ & & & & & & \downarrow & & & & \\ 0 & & 0 & & 0 & & 1 & \rightarrow & -0.667 & & -0.333 \\ & & & & & & & & \downarrow & & \\ 0 & & 0 & & 0 & & 0 & & 1 & \rightarrow & -0.999 \\ & & & & & & & & & & \\ 0 & & 0 & & 0 & & 0 & & 0 & & 0 \end{bmatrix} \tag{5.7}$$

The nonzero elements (i, i) and (i, j) indicate the existence of equations (measurements) that relate the state variables corresponding to their columns, i.e., they indicate the possibility of estimating the power flows between the buses i and j through the available measurements. Following this criterion and remembering that the state variables of the linear estimator are the phase angles of the nodal voltages, the following relations (or connections between the state variables) are verified:

- Row 1: elements (1,1) and (1,2): θ_1 connected with θ_2;
- Row 2: elements (2,2) and (2,3): θ_2 connected with θ_3;
- Row 3: elements (3,3) and (3,4): θ_3 connected with θ_4;
- Row 4: elements (4,4) and (4,5): θ_4 connected with θ_5;
- Row 5: elements (5,5) and (5,6): θ_5 connected with θ_6.

In the context of observability analysis, the existence of a single connected factorization path indicates that it is sufficient to adopt one of the buses of the EPS as an angular reference in order to make it possible to estimate the voltage phase angles of all other buses of the EPS.

The union of the factorization paths, associated with the factored matrix illustrated above, can be graphically shown through the following graph tree (defined as a factorization path tree):

Each factorization path is obtained through the following procedure (Tinney et al., 1985): the row (i) of the matrix U is traversed from the element (i, i) to the first nonzero element (i, j) and passes to the row

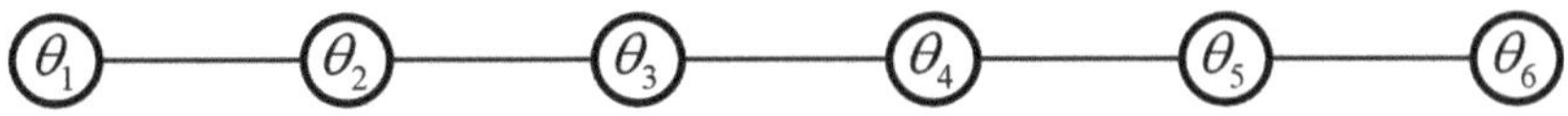

Fig. 5.6 Factorization paths tree.

corresponding to the column of that element (row (j)). The process is repeated until all rows of the U-matrix have been traversed, except for the last row.

Because the factored matrix G_θ has rank equal to $(n-1)$, its triangular factorization resulted in only one null pivot, in the last diagonal, and all the resulting factorization paths have a connection, giving rise to a single factorization path (Bretas, 1996) (represented by the factorization path tree shown in Fig. 5.6).

In the case of a nonobservable EPS, the corresponding matrix G_θ will have rank lower than $(n-1)$ and thus, as mentioned previously, during the factorization of this matrix will appear a null pivot on diagonal i, where $i<n$, indicating that the system is not observable as a whole, and the other elements of row and column i will be null, as shown in Fig. 5.4. In terms of factorization paths, this means that the remaining phase angles, corresponding to the columns of $i+1$ through n, will be part of other (at least one more) factorization paths that will have no connection to the previous factorization path.

To obtain all the factorization paths associated with the triangular factorization of the G_θ matrix associated with an unobservable EPS, every time a null pivot is found, the triangular factorization process must continue from the next diagonal, provided that it is not the diagonal n, for which the triangular factorization process ends.

The existence of disconnected factorization paths demonstrates the lack of measurements relating the state variables of these factorization paths. Thus through these factorization paths it is possible to perform the procedures of restoration of observability, or identification of islands, as follows.

Identification of observable islands

1. If there is no available power injection measurement relating nodes (state variables) of different factorization paths, the system is not observable as a whole and the nodes belonging to each of the n_{cf} factorization paths will constitute n_{cf} observable subnetworks, i.e., n_{cf} observable islands.
2. If the available measurement set has at least one injection measurement relating nodes of different factorization paths, the system is not observable as a whole and it is not possible to ensure that the subnetworks

associated with each isolated factorization path constitute observable islands. To find the observable islands, that injection measurement must be identified, and discarded, in order to obtain a new G_θ matrix (it is an irrelevant injection measurement). Then, the triangular factorization of this matrix must be performed. The process of discarding measurements and triangular factorization ends when there is no further injection measurement relating nodes of different factorization paths. Thus after the process of identifying and discarding irrelevant measurements, the nodes belonging to each of the factorization paths found will constitute n_{cf} observable subnetworks, i.e., n_{cf} observable islands.

Restoration of observability

To make the system observable, it is enough to add pseudo measurements of flow or power injection that relate state variables belonging to distinct factorization paths, connecting them to each other (one measurement at a time). In this way, these factorization paths connected through these pseudo measurements become a single factorization path, making the entire system observable.

Algorithm:

- **Step 1:** With the available measurement set, assemble the Gain matrix G_θ (considering $R^{-1}=I$ and $x_l=1$ pu, $l=1, \ldots, n_l$, where n_l is the total number of branches of the network in analysis).
- **Step 2:** Perform the triangular factorization of the matrix G_θ.
- **Step 3:** Identify the factorization paths associated with the factorization of the G_θ matrix. If the triangular factorization of the matrix G_θ resulted in only one factorization path, then the network is completely observable, and the process can stop. Otherwise, the network is not observable as a whole. For Identification of Observable Islands, go to Step 4 and for Observability Restoration go to Step 6.
- **Step 4:** If there are no power injection measurements relating nodes of different factorization paths, then the associated subnetworks with each isolated factorization path are already observable islands, and the process can stop. Otherwise, go to Step 5.
- **Step 5:** Identify and discard the power injection measurements relating nodes of different factorization paths. Update the G_θ matrix and redo its triangular factorization. Check the factorization paths and return to Step 4.
- **Step 6:** Select a flow or power injection pseudo measurement that relates state variables of different factorization paths.

- **Step 7:** Add the selected pseudo measurement in the previous step to the original measurement set and update the G_θ matrix. Then, redo the triangular factorization of this matrix and identify the factorization paths. If this factorization results in only one factorization path, then the network becomes observable as a whole, stop. Otherwise, return to Step 6.

Example 5.1a: To exemplify the application of the algorithm for observability analysis based on the triangular factorization of the matrix G_θ and in factorization paths, consider the six-bus EPS associated with the measurement set illustrated in Fig. 5.7.

Step 1: With the measurement set available, obtain the Gain matrix G_θ (considering $R^{-1} = I$ and $x_l = 1$ pu, $l = 1, \ldots, n_l$):

$$G_\theta = \begin{bmatrix} 5 & -3 & -2 & 0 & 0 & 0 \\ -3 & 2 & 1 & 0 & 0 & 0 \\ -2 & 1 & 2 & -3 & 1 & 1 \\ 0 & 0 & -3 & 10 & -4 & -3 \\ 0 & 0 & 1 & -4 & 2 & 1 \\ 0 & 0 & 1 & -3 & 1 & 1 \end{bmatrix} \tag{5.8}$$

Step 2: Through the triangular factorization of this matrix, we obtain the following matrix of factors:

$$U_{G\theta} = \begin{bmatrix} 1 & \rightarrow & -0.6 & & -0.4 & & 0 & & 0 & 0 \\ & & \downarrow & & & & & & & \\ 0 & & 1 & \rightarrow & -0.999 & & 0 & & 0 & 0 \\ & & & & \downarrow & & & & & \\ 0 & & 0 & & 1 & \rightarrow & -3 & & 1 & 1 \\ & & & & & & \downarrow & & & \\ 0 & & 0 & & 0 & & 1 & \rightarrow & 1 & 0 \\ & & & & & & & & \downarrow & \\ 0 & & 0 & & 0 & & 0 & & 0 & 0 \\ & & & & & & & & & \\ 0 & & 0 & & 0 & & 0 & & 0 & 0 \end{bmatrix} \tag{5.9}$$

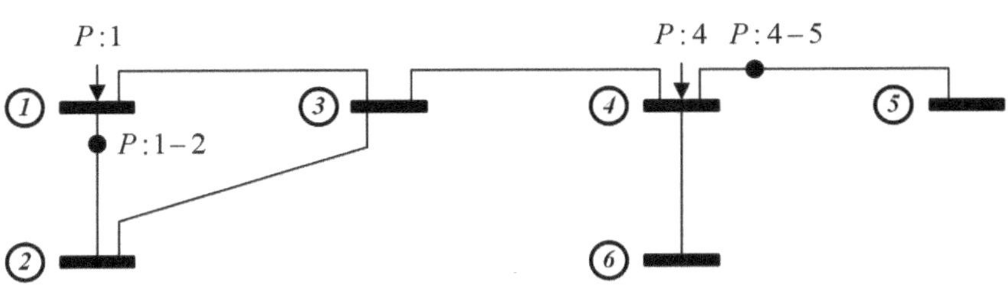

Fig. 5.7 Six-bus power system associated with a measurement set (•, flow measurements; →, injection measurements).

Step 3: According to the arrows indicated in the factor matrix $U_{G\theta}$, the factorization resulted in two factorization paths, which can be represented by two factorization path trees:

- Factorization path tree [1]: θ1–θ2–θ3–θ4–θ5;
- Factorization path tree [2]: θ6.

In this way, the EPS is not completely observable. Initially choosing the option "Identification of Observable Islands", we will go to Step 4 of the algorithm.

Identification of observable Islands

Step 4: As the injection measurement P:4 relates state variables of distinct factorization paths (the state variables θ3, θ4, and θ5 belong to the factorization path tree [1] and the variable θ6 belongs to the tree [2]), go to Step 5.

Step 5: The injection measurement P:4 is discarded from the measurement set, the matrix H_{DC} is updated and the matrix G_θ is recalculated:

$$G_\theta = \begin{bmatrix} 5 & -3 & -2 & 0 & 0 & 0 \\ -3 & 2 & 1 & 0 & 0 & 0 \\ -2 & 1 & 1 & 0 & 0 & 0 \\ 0 & 0 & 0 & 1 & -1 & 0 \\ 0 & 0 & 0 & -1 & 1 & 0 \\ 0 & 0 & 0 & 0 & 0 & 0 \end{bmatrix} \tag{5.10}$$

Through the triangular factorization of this matrix, the following matrix of factors is obtained:

$$U_{G\theta} = \begin{bmatrix} 1 & \rightarrow & -0.6 & & -0.4 & 0 & & 0 & 0 \\ & & \downarrow & & & & & & \\ 0 & & 1 & \rightarrow & -0.999 & 0 & & 0 & 0 \\ 0 & & 0 & & 0 & 0 & & 0 & 0 \\ 0 & & 0 & & 0 & 1 & \rightarrow & -1 & 0 \\ 0 & & 0 & & 0 & 0 & & 0 & 0 \\ 0 & & 0 & & 0 & 0 & & 0 & 0 \end{bmatrix} \tag{5.11}$$

According to the arrows indicated in the factor matrix UG_θ, factorization resulted in three factorization path trees:

- Factorization path tree [1]: θ1–θ2–θ3;
- Factorization Path Tree [2]: θ4–θ5;
- Factorization Path Tree [3]: θ6.

Return to **Step 4**.

Step 4: As there is no more injection measurement relating state variables of distinct factorization paths, the associated subnetworks, associated with each isolated factorization path trees, constitute the observable islands of the network. **End of algorithm**.

Returning now to Step 3 of the algorithm, but now choosing the Observability Restoration option, go to Step 6. Remembering that at this point the factorization path trees to be considered are the two resulting from the original G_θ matrix factorization, the matrix obtained considering the availability of all the measurements illustrated in Fig. 5.7. These trees are as follows:

- Factorization path tree [1]: θ1–θ2–θ3–θ4–θ5;
- Factorization path tree [2]: θ6.

Observability restoration

To exemplify the process of restoration of observability, the following pseudo measurements will be considered: Ps: 2 (active power injection at bus 2), Ps: 1–3 (active power flow from bus 1 to bus 3), and Ps: 6 (active power injection at bus 6).

Step 6: Among the pseudo measurements available for restoration of observability, the pseudo injection in bus 6 (Ps: 6) is the only one that relates state variables of the two factorization path trees. Thus this measurement is selected to restore observability.

Step 7: Adding the pseudo measurement Ps: 6 to the original set of measurements, the following matrix G_θ is obtained:

$$G_\theta = \begin{bmatrix} 5 & -3 & -2 & 0 & 0 & 0 \\ -3 & 2 & 1 & 0 & 0 & 0 \\ -2 & 1 & 2 & -3 & 1 & 1 \\ 0 & 0 & -3 & 11 & -4 & -4 \\ 0 & 0 & 1 & -4 & 2 & 1 \\ 0 & 0 & 1 & -4 & 1 & 2 \end{bmatrix} \tag{5.12}$$

Through the triangular factorization of this matrix, the following matrix of factors is obtained:

$$U_{G\theta} = \begin{bmatrix} 1 & \rightarrow & -0.6 & & -0.4 & & 0 & & 0 & & 0 \\ & & \downarrow & & & & & & & & \\ 0 & & 1 & \rightarrow & -0.999 & & 0 & & 0 & & 0 \\ & & & & \downarrow & & & & & & \\ 0 & & 0 & & 1 & \rightarrow & -3 & & 1 & & 1 \\ & & & & & & \downarrow & & & & \\ 0 & & 0 & & 0 & & 1 & \rightarrow & -0.5 & & -0.5 \\ & & & & & & & & \downarrow & & \\ 0 & & 0 & & 0 & & 0 & & 1 & \rightarrow & -1 \\ & & & & & & & & & & \\ 0 & & 0 & & 0 & & 0 & & 0 & & 0 \end{bmatrix} \tag{5.13}$$

According to the arrows indicated in the matrix of the factors UG_θ, the factorization resulted in only one tree of factorization paths:

- Factorization path tree [1]: θ1–θ2–θ3–θ4–θ5–θ6.

As we obtained only one factorization path tree, the system became observable with the inclusion of the pseudo measurement Ps: 6. **End of the algorithm**.

Note 5.4: The method for observability analysis proposed in Bretas (1996) was presented in this section considering the triangular factorization of the matrix G_θ, the Gain matrix of the linear WLS state estimator. This is because this method, as well as of the method proposed by Wu and Monticelli (1985), has been developed based on this matrix. It is important to remember that for metering systems in which active and reactive power measurements are performed in pairs, the algebraic observability of the linear model (which is equivalent to the observability of the $P\theta$ model) guarantees QV observability, provided that there is at least one measurement of nodal voltage magnitude. It should be noted, however, that the method proposed in Bretas (1996) can also be applied to the Gain matrix of the nonlinear WLS state estimator, since the method is based on the analysis of the linear dependence relations of the rows of the Gain matrix. In the same way, the method for observability analysis based on the triangular factorization of the Jacobian matrix of the WLS estimator will be presented in the next section considering the Jacobian matrix of the linear WLS estimator. However, it can be used in the Jacobian matrix of the nonlinear WLS estimator. In this sense, it is emphasized that in Toyoshima et al. (2012) this method was employed considering the Jacobian matrix of the three-phase version of the nonlinear WLS state estimator.

Method based on the triangular factorization of the Jacobian matrix and the concept of factorization paths

In order to verify if the system under analysis is observable as a whole, according to the method proposed in Benedito et al. (2009), the triangular factorization of the corresponding matrix H_{DC}, the Jacobian matrix of the linear WLS state estimator, is performed. It is worth noting that this factorization must be performed through linear combinations of the H_{DC} matrix columns, as presented in London et al. (2007), which correspond to the state variables of the system. Thus if the system is observable, after the complete factorization of the H_{DC} matrix, the matrix H_Δ will be obtained, which relates the measurements with the equivalent state variables, which are linear combinations of the system state variables.

It will be shown in the next section that the H_Δ matrix enables the analysis of the local redundancy of measurements. To obtain it, complete factorization of the H_{DC} matrix is required, as detailed in London et al. (2007). However, for observability analysis, only the partial matrix factorization is required, i.e., the elimination of the upper triangle of the H_{DC} matrix, which will be presented next.

If the system under analysis is $P\theta$ observable, the triangular factorization of the corresponding matrix H_{DC} will result in only one null pivot in the diagonal (n, n) (London et al., 2007), where n is the number of buses of the analyzed system. In this case, the last column of the resulting matrix will consist only of zeros. It is worth remembering that, during the factorization process, row permutations may be necessary to avoid possible null pivots.

However, if the system is not observable, a null pivot will appear on the diagonal element (i, i) during triangular factorization, where $i < n$. In this case, there will be no nonzero element in the other rows of the H_{DC} matrix that is being factorized in the null pivot column, indicating the lack of measurements providing information of the equivalent state variable corresponding to that column.

When the system is not observable as a whole, the method proposed in Benedito et al. (2009) allows:

- Restoration of observability if the necessary pseudo measurements are available;
- Otherwise, identification of observable islands.

It is important to highlight that the process of restoration of observability through the factorization of the H_{DC} matrix is simple and straightforward (not iterative), without need for identification of the unobservable branches and the factorization paths, nor the solution of systems of algebraic equations.

Restoration of observability

In order to restore observability, a pseudo measurement search is performed to provide information about the equivalent state variable corresponding to the null pivot column. This is done through the triangular factors obtained during the H_{DC} matrix factorization process. This search can be summarized in the following steps (Benedito et al., 2009):

1. Insert a new row in the H_{DC} matrix being factorized, where the equation associated with the first available pseudo measurement will be stored.
2. Apply the triangular factors to this new row.

3. If a nonzero element appears in the null pivot column in the new row, it is because the corresponding pseudo measurement provides the necessary information to restore observability.

It should be noted that, through the referred procedure, it is guaranteed that the restoration of observability will be carried out through critical pseudo measurements. If it is not possible to restore observability through the available pseudo measurements, the methodology proposed in Benedito et al. (2009) will allow the identification of the observable islands, but for this it will use the factorization paths associated with triangular factorization of the corresponding matrix H_{DC}, as will be shown next.

Identification of observable islands

The identification of observable islands is performed through the analysis of the factorization paths associated with the triangular factorization of the H_{DC} matrix, without the requirement of any solution of systems of algebraic equations.

The theoretical basis of the procedure for identifying observable islands through the graph paths associated with the triangular factorization of the H_{DC} matrix was presented in Benedito et al. (2009).

In order to obtain the factorization paths, it is necessary to include a new row in the matrix that is being factorized with only one nonzero element equal to 1 in the null pivot column and continue the factorization until the diagonal (n, n) is reached. Note that this procedure is equivalent to the insertion of a critical pseudo measurement of voltage phase angle, providing only information of the unobservable equivalent state variable, i.e., of the equivalent state variable corresponding to the null pivot column (Benedito et al., 2009; Wu and Monticelli, 1985; Monticelli and Wu, 1985). The insertion of pseudo measurements of voltage phase angle has the objective of making the system "artificially observable", making the rank of the matrix H_{DC} equal to n, thus making it possible to obtain of the factorization paths.

Fig. 5.8 shows the matrices L and U of the Lu factorization (triangular factorization) of the H_{DC} matrix associated with an unobservable system, which contains m active power measurements, after the inclusion of a pseudo measurement of angle (θ_1^p). The blank areas of these figures correspond to null elements and the shaded areas represent possibly nonzero elements.

The methodology proposed in Benedito et al. (2009) allows the identification of the observable islands through the following steps:

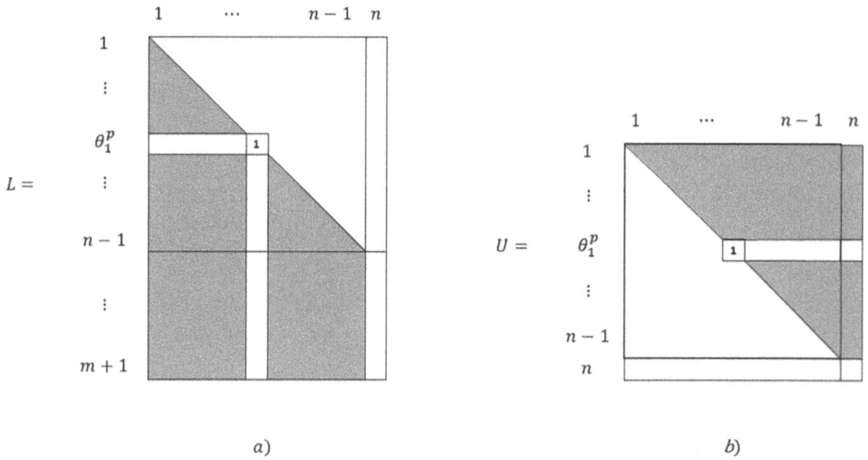

Fig. 5.8 Decomposition LU of the H_{DC} matrix associated with an unobservable system: (A) lower trapezoidal matrix L and (B) upper matrix with triangular factors U.

1. If there is no available power injection measurement relating nodes (or equivalent state variables) of different factorization paths in the available measurement set, the system is not observable as a whole and the nodes belonging to each of the n_{cf} factorization paths found will constitute n_{cf} observable subnetworks, i.e., n_{cf} observable islands.
2. If in the available measurement set there is at least one injection measurement relating nodes of different factorization paths, the system is not observable as a whole and it is not possible to ensure that the subsystems associated with each isolated factorization path constitute observable islands. To find the observable islands, that measurement must be identified and discarded, obtaining a new H_{DC} matrix (it is an irrelevant injection measurement). Then, redo its triangular factorization. The process of discarding measurements and factorization ends when there is not injection measurement relating to nodes of different factorization paths anymore. Thus after the process of identifying and discarding irrelevant measurements, the nodes belonging to each of the factorization paths found will constitute n_{cf} observable subsystems, i.e., n_{cf} observable islands.

Note 5.5: The process of identifying observable islands can become iterative if the case described in Item 2 occurs (in the same way as the island identification process of the method based on the triangular factorization of the Gain matrix). This is because when irrelevant measurements are identified and discarded, other irrelevant measurements may appear.

Note 5.6: The factor-based method of Jacobian matrix H_{DC} can be executed by applying the triangular factorization process directly in this matrix or in the transposed Jacobian matrix H_{DC}^T. It is worth remembering that when applied to the H_{DC} matrix, the process must occur by combining the columns of this matrix that correspond to the system state variables. Thus when applying the factorization in the transposed Jacobian matrix H_{DC}^T, this process must combine only the rows of this matrix that correspond to the state variables of the system. As discussed in London et al. (2007), considering the structure of the H_{DC}^T matrix, it becomes easier to apply the factorization process to this matrix rather than directly applying it to H_{DC}. Due to this ease of implementation of the factorization process, we chose to present the algorithm using the H_{DC}^T matrix. It should be noted that the H_{DC}^T matrix has the same properties as the H_{DC} matrix. Thus everything presented in relation to the H_{DC} matrix will apply to the H_{DC}^T matrix, provided that the differences between these matrices are observed, due to the fact that one is the transpose of the other.

Algorithm:

- **Step 1:** With the measurement set available, obtain the transposed Jacobian matrix H_{DC}^T (considering $x_l = 1$ pu, $l=1, \ldots, n_l$, where n_l is the total number of branches of the electric grid under analysis). If $m<n$, introduce $m-n$ new columns into the H_{DC}^T matrix formed by only zeros, where m and n are, respectively, the numbers of available measurements and buses of the system under analysis.
- **Step 2**: Perform the triangular factorization of the H_{DC}^T matrix, storing the triangular factors, performing permutations of columns when possible to avoid null pivots, until $H_F^T(d,d)=0$. If $d=n$, then the system is observable, *stop* (when the system is observable, by permutation of columns it will always be possible to finalize the factorization on the diagonal (n,n)). Otherwise, the system is not observable, and any column of the matrix being factorized will have a nonzero element on row d. For Identification of Observable Islands, go to Step 3 and for Observability Restoration go to Step 6.
- **Step 3**: Enter a voltage phase angle pseudo measurement, providing information of the unobservable equivalent state variable corresponding to row d (which is equivalent to inserting a new column in the matrix being factorized with only one nonzero

element in row d) and continue the factorization process until the diagonal $d = n$. Note that this process may require column permutations to avoid null pivots or the inclusion of other voltage phase angle pseudo measurements. Proceed to the next step.

- **Step 4:** Find the factorization paths through the elements of the corresponding matrix of factors U. Then check if there is any injection measurement relating variables (nodes) of different factorization paths. If it does not exist, then the associated subnetworks, with every factorization path isolated, are already observable islands, *stop*. Otherwise, go to Step 5.
- **Step 5:** Identify and discard the power injection measurements relating nodes of different factorization paths. Update the H_{DC}^T matrix, redo the triangular factorization (as described in Step 3) and return to Step 4.
- **Step 6:** Select the first available pseudo measurement (it is recommended to the select the pseudo measurement considered of best quality or one related to the bus that corresponds to the equivalent state variable d). Go to the next step.
- **Step 7:** Create a new column in the matrix H_F^T under factorization to store the equation corresponding to the selected pseudo measurement. Then apply to this column, the triangular factors obtained so far in the factorization process. If after the application of the triangular factors a nonzero element appears on the null pivot row (row d), it is because the selected pseudo measurement provides the information of the equivalent state variable corresponding to row d, go to the next step. Otherwise, go to Step 9.
- **Step 8:** Continue the triangular factorization of the H_F^T matrix updated in Step 7, performing column permutations when possible to avoid null pivots, until $H_F^T(d, d) = 0$. If $d = n$, the system became observable considering the pseudo measurements that were selected, *stop*. Otherwise, go to Step 9.
- **Step 9:** Select the next available pseudo measurement and return to Step 7.

Example 5.1b: To exemplify the application of the algorithm for observability analysis based on the triangular factorization of the H_{DC}^T matrix and on factorization paths, the six-bus EPS associated with the set of measurements shown in Fig. 5.7 is analyzed.

Step 1: With the available measurement set, obtain the transposed Jacobian matrix H_{DC}^T (considering $x_l = 1$ $p.$ $u.$, $l = 1, \ldots, n_l$):

Continued

$$H_{DC}^{T} = \begin{array}{c} \\ \theta_1 \\ \theta_2 \\ \theta_3 \\ \theta_4 \\ \theta_5 \\ \theta_6 \end{array} \begin{array}{c} \begin{array}{cccccc} P{:}1\text{-}2 & P{:}1 & P{:}4 & P{:}4\text{-}5 & & \end{array} \\ \left[\begin{array}{cccccc} 1 & 2 & 0 & 0 & 0 & 0 \\ -1 & -1 & 0 & 0 & 0 & 0 \\ 0 & -1 & -1 & 0 & 0 & 0 \\ 0 & 0 & 3 & 1 & 0 & 0 \\ 0 & 0 & -1 & -1 & 0 & 0 \\ 0 & 0 & -1 & 0 & 0 & 0 \end{array}\right] \end{array}$$

Notice that two columns formed by zeros were included in the H_{DC}^{T} matrix because only four measurements were available for the six-bus system under analysis $(6-4=2)$.

Step 2: Performing the triangular factorization of this matrix, we obtain (note that in some situations null pivots can be avoided with the permutation of columns):

$$H_{F}^{T} = \begin{array}{c} \\ \theta_1 \\ \theta_2 \\ \theta_3 \\ \theta_4 \\ \theta_5 \\ \theta_6 \end{array} \begin{array}{c} \begin{array}{cccccc} P{:}1\text{-}2 & P{:}1 & P{:}4 & P{:}4\text{-}5 & & \end{array} \\ \left[\begin{array}{cccccc} 1 & 2 & 0 & 0 & 0 & 0 \\ (1) & 1 & 0 & 0 & 0 & 0 \\ 0 & (1) & -1 & 0 & 0 & 0 \\ 0 & 0 & (3) & 1 & 0 & 0 \\ 0 & 0 & (-1) & (1) & 0 & 0 \\ 0 & 0 & (-1) & 0 & 0 & 0 \end{array}\right] \end{array}$$

The elements that appear in parentheses are the triangular factors, that is, the elements of the corresponding matrix of the factors.

Since $H_F^T(5,5)=0$, and $n \neq 5$ $(n=6)$, EPS is not observable. Choosing the option Identification of Observable Islands, we will go to Step 3 of the algorithm.

Identification of observable islands

Step 3: Two new columns are created in the matrix being factored, with nonzero elements, respectively, in rows $d=5$ and $d=6$ (addition of pseudo measurements of voltage phase angles $P_{\theta 5}$ and $P_{\theta 6}$). These will be the new columns "5" and "6" of that matrix and the other columns must be shifted, resulting in the following matrix:

$$
H_F^T = \begin{array}{c} \\ \theta_1 \\ \theta_2 \\ \theta_3 \\ \theta_4 \\ \theta_5 \\ \theta_6 \end{array}
\begin{array}{c}
\begin{array}{cccccccc} \text{P:1-2} & \text{P:1} & \text{P:4} & \text{P:4-5} & P_{\theta 5} & P_{\theta 6} & & \end{array} \\
\left[\begin{array}{cccccccc}
1 & 2 & 0 & 0 & 0 & 0 & 0 & 0 \\
(1) & 1 & 0 & 0 & 0 & 0 & 0 & 0 \\
0 & (1) & -1 & 0 & 0 & 0 & 0 & 0 \\
0 & 0 & (3) & 1 & 0 & 0 & 0 & 0 \\
0 & 0 & (-1) & (1) & 1 & 0 & 0 & 0 \\
0 & 0 & (-1) & 0 & 0 & 1 & 0 & 0
\end{array}\right]
\end{array}
$$

Step 4: Through the triangular factors, there are two factorization paths, which can be represented by the following two factorization path trees:

$$
H_F^T = \begin{array}{c} \\ \theta_1 \\ \\ \theta_2 \\ \\ \theta_3 \\ \\ \theta_4 \\ \\ \theta_5 \\ \\ \theta_6 \end{array}
\begin{array}{c}
\begin{array}{cccccccc} \text{P:1-2} & \text{P:1} & \text{P:4} & \text{P:4-5} & P_{\theta 5} & P_{\theta 6} & & \end{array} \\
\left[\begin{array}{cccccccc}
1 & 2 & 0 & 0 & 0 & 0 & 0 & 0 \\
\downarrow & & & & & & & \\
(1) \rightarrow & 1 & 0 & 0 & 0 & 0 & 0 & 0 \\
 & \downarrow & & & & & & \\
0 & (1) \rightarrow & -1 & 0 & 0 & 0 & 0 & 0 \\
 & & \downarrow & & & & & \\
0 & 0 & (3) \rightarrow & 1 & 0 & 0 & 0 & 0 \\
 & & & \downarrow & & & & \\
0 & 0 & (-1) & (1) \rightarrow & 1 & 0 & 0 & 0 \\
 & & & & & & & \\
0 & 0 & (-1) & 0 & 0 & 1 & 0 & 0
\end{array}\right]
\end{array}
$$

- Factorization path tree [1]: θ_1 - θ_2 - θ_3 - θ_4 - θ_5;
- Factorization path tree [2]: θ_6.

As the injection measurement $P{:}4$ relates state variables of distinct factorization paths (the state variables θ_3, θ_4, and θ_5 belong to the factorization path tree [1] and the variable θ_6 belongs to the tree [2]), go for Step 5.

Step 5: Injection measurement $P{:}4$ is discarded from the measurement set, the H_{DC}^T matrix is updated, and the triangular factorization process is reapplied, as described in Step 3, resulting in (observe that it was necessary to insert a voltage phase angle pseudo measurement $P_{\theta 3}$):

Continued

$$
H_F^T = \begin{array}{c} \\ \theta_1 \\ \\ \theta_2 \\ \\ \theta_3 \\ \\ \theta_4 \\ \\ \theta_5 \\ \\ \theta_6 \end{array}
\begin{array}{cccccccc}
\text{P:1-2} & \text{P:1} & P_{\theta 3} & \text{P:4-5} & P_{\theta 5} & P_{\theta 6} & & \\
1 & 2 & 0 & 0 & 0 & 0 & 0 & 0 \\
\downarrow & & & & & & & \\
(1) \rightarrow & 1 & 0 & 0 & 0 & 0 & 0 & 0 \\
 & \downarrow & & & & & & \\
0 & (1) \rightarrow & 1 & 0 & 0 & 0 & 0 & 0 \\
 & & & & & & & \\
0 & 0 & 0 & 1 & 0 & 0 & 0 & 0 \\
 & & & \downarrow & & & & \\
0 & 0 & 0 & (1) \rightarrow & 1 & 0 & 0 & 0 \\
 & & & & & & & \\
0 & 0 & 0 & 0 & 0 & 1 & 0 & 0
\end{array}
$$

Return to **Step 4**.

Step 4: Through the triangular factors, there are three factorization paths, which can be represented by the following factorization path trees:

- Factorization path tree [1]: $\theta_1 - \theta_2 - \theta_3$;
- Factorization path tree [2]: $\theta_4 - \theta_5$;
- Factorization path tree [3]: θ_6.

As there is no further injection measurement relating state variables of distinct factorization paths, the associated subnetworks with each isolated factorization path tree constitute the observable islands of the network. **End of algorithm.**

Returning now to Step 2 of the algorithm, but now choosing the Observability Restoration option, go to Step 6.

Observability restoration

To exemplify the process of restoration of observability, as in Example 5.1a, the following pseudo measurements will be considered: $Ps\!:\!2$, $Ps\!:\!1-3$, and $Ps\!:\!6$.

Step 6: Selecting the first pseudo measurement available for analysis ($Ps\!:\!2$).

Step 7: Create a new column in the H_F^T matrix, which is under factorization, to store the equation of the pseudo measurement $Ps\!:\!2$. Then, in this column, apply the triangular factors obtained previously in the factorization process. The H_F^T matrix becomes:

$$H_F^T = \begin{array}{c} \\ \theta_1 \\ \theta_2 \\ \theta_3 \\ \theta_4 \\ \theta_5 \\ \theta_6 \end{array} \begin{array}{ccccccc} \text{P:1-2} & \text{P:1} & \text{P:4} & \text{P:4-5} & & & \text{Ps:2} \\ 1 & 2 & 0 & 0 & 0 & 0 & -1 \\ (1) & 1 & 0 & 0 & 0 & 0 & 1 \\ 0 & (1) & -1 & 0 & 0 & 0 & 0 \\ 0 & 0 & (3) & 1 & 0 & 0 & 0 \\ 0 & 0 & (-1) & (1) & 0 & 0 & 0 \\ 0 & 0 & (-1) & 0 & 0 & 0 & 0 \end{array}$$

As $H_F^T(5,7)=0$, $Ps:2$ does not provide the information of the equivalent state variable corresponding to row $d=5$. Go to Step 9.

Step 9: The pseudo measurement $Ps:1-3$ is selected. Return to Step 7.

Step 7: Create a new column in the H_F^T matrix to store the pseudo measurement equation $Ps:1-3$. Then, in this column, the triangular factors previously obtained in the factorization process are applied. The H_F^T matrix becomes:

$$H_F^T = \begin{array}{c} \\ \theta_1 \\ \theta_2 \\ \theta_3 \\ \theta_4 \\ \theta_5 \\ \theta_6 \end{array} \begin{array}{ccccccc} \text{P:1-2} & \text{P:1} & \text{P:4} & \text{P:4-5} & & & \text{Ps:1-3} \\ 1 & 2 & 0 & 0 & 0 & 0 & 1 \\ (1) & 1 & 0 & 0 & 0 & 0 & 1 \\ 0 & (1) & -1 & 0 & 0 & 0 & 0 \\ 0 & 0 & (3) & 1 & 0 & 0 & 0 \\ 0 & 0 & (-1) & (1) & 0 & 0 & 0 \\ 0 & 0 & (-1) & 0 & 0 & 0 & 0 \end{array}$$

As $H_F^T(5,7)=0$, Ps: 1–3 does not provide the information of the equivalent state variable corresponding to row $d=5$. Go to Step 9.

Step 9: The pseudo measurement $Ps:6$ is selected. Return to Step 7.

Step 7: Create a new column in the H_F^T matrix to store the pseudomeasurement equation $Ps:6$. Then, in this column, apply the triangular factors obtained in the factorization process. The H_F^T matrix becomes:

$$H_F^T = \begin{array}{c} \\ \theta_1 \\ \theta_2 \\ \theta_3 \\ \theta_4 \\ \theta_5 \\ \theta_6 \end{array} \begin{array}{ccccccc} \text{P:1-2} & \text{P:1} & \text{P:4} & \text{P:4-5} & & & \text{Ps:6} \\ 1 & 2 & 0 & 0 & 0 & 0 & 0 \\ (1) & 1 & 0 & 0 & 0 & 0 & 0 \\ 0 & (1) & -1 & 0 & 0 & 0 & 0 \\ 0 & 0 & (3) & 1 & 0 & 0 & -1 \\ 0 & 0 & (-1) & (1) & 0 & 0 & -1 \\ 0 & 0 & (-1) & 0 & 0 & 0 & 1 \end{array}$$

As $H_F^T(5,7)=-1$, $Ps:6$ provides the information of the equivalent state variable corresponding to row $d=5$. Go to Step 8.

Continued

Step 8: Continue the triangular factorization of the H_F^T matrix updated in Step 7, performing column permutations when possible to avoid null pivots. You get:

$$H_F^T = \begin{array}{c} \\ \theta_1 \\ \theta_2 \\ \theta_3 \\ \theta_4 \\ \theta_5 \\ \theta_6 \end{array} \begin{array}{c} \begin{array}{ccccccc} P{:}1\text{-}2 & P{:}1 & P{:}4 & P{:}4\text{-}5 & Ps{:}6 & & \end{array} \\ \left[\begin{array}{ccccccc} 1 & 2 & 0 & 0 & 0 & 0 & 0 \\ (1) & 1 & 0 & 0 & 0 & 0 & 0 \\ 0 & (1) & -1 & 0 & 0 & 0 & 0 \\ 0 & 0 & (3) & 1 & -1 & 0 & 0 \\ 0 & 0 & (-1) & (1) & -1 & 0 & 0 \\ 0 & 0 & (-1) & 0 & (1) & 0 & 0 \end{array}\right] \end{array}$$

Since $H_F^T(6,6) = 0$, and $n = 6$, EPS became observable with the inclusion of the pseudo measurement $Ps: 6$. **End of the algorithm.**

5.2 Redundancy of analog measurements

5.2.1 Introduction

In terms of redundancy of analog measurements for PSSE, the following definitions are of paramount importance:

Global redundancy

Global redundancy of a measurement set is the ratio between the number of measurements available and the number of state variables to be estimated.

Critical measurement

Critical measurement (CM) is the measurement that, if removed from the measurement set of an observable system (or an observable island), makes that same system (or island) unobservable.

Noncritical measurements are called redundant measurements and only such measurements can have nonzero residuals (Clements et al., 1981).

Critical set

A critical set of measurements, also known in the literature as minimally dependent sets of measurements or bad data groups, can be defined, according to Ayres and Haley (1986), in two ways:

Numerical definition

Critical sets of measurements (CSMs) are those corresponding to the submatrices of the covariance matrix of residuals[a] with rank equal to 1.

Topological definition

A CSM is the set of measurements formed by noncritical measurements, in which the elimination of any measurement makes the other measurements critical.

It should be noted that global redundancy does not accurately express the redundancy level of measurements in PSSE. Therefore even for metering systems with a high level of global redundancy (greater than 3.5), there is the possibility of CMs or CSMs (London et al., 2007; Celik and Liu, 1995).

In Al-Atwan and Koglin (1998), the problem of local redundancy of measurements for PSSE effect was discussed in detail. Among the two concepts of local redundancy presented and discussed, the most important was the concept of integer local redundancy. According to this reference, in order to obtain the entire local redundancy level of a measurement i, measurements whose residuals are strongly correlated with the residual of that measurement must be eliminated to the extent i becomes critical. Thus if measurement i becomes a CM with the elimination of only one of those measurements, measurement i has integer local redundancy equal to 1.

According to this definition, CM has an integer local redundancy, which we will call simply the local redundancy level, equal to 0. On the other hand, the measurements belonging to the CSMs have integer local redundancy of 1.

Another way to understand the concept of local redundancy of measurements is to relate redundancy and observability problems. In this context, since the removal of a CM causes the loss of observability, the CM has a local redundancy level of 0, since it is the only measurement giving information about a state variable. A redundant measurement has local redundancy level equal to 1, if removing it does not cause loss of observability, but instead results in additional measurements with local redundancy level equal to 0, i.e., of CMs. Thus to cause loss of observability, it is necessary to remove at least two redundant measurements with a local redundancy level of 1. In more general terms, a redundant measurement has local redundancy level equal to k if its suppression does not cause the loss of observability but instead results in additional measurements with local redundancy level $k-1$. Considering the above, in order to cause the loss of observability, it is necessary to

[a] The covariance matrix of residuals will be presented in Chapter 6.

remove at least $k+1$ redundant measurements with local redundancy level equal to k.

To illustrate the development of methods for treating redundancy of measurements for PSSE, the following section highlights some of the available methods, including the procedures used, advantages, and disadvantages.

5.2.2 Methods developed

The methods designed to identify CMs and CSMs, like those developed for observability analysis, can be classified, in general, as topological or numerical. The methods belonging to the first group use the concept of topological observability, presenting a combinatorial nature, not requiring numerical computation, requiring only the topology of the network, as well as the type and localization of the available measurements (Clements et al., 1981; Crainic et al., 1990; Costa et al., 1990; Bretas et al., 2005).

In Clements et al. (1981), the first method to identify CMs was developed. The theorems relative to CM presented in this work served as a basis for other methods. The technique is based on the observable spanning tree, which can identify as CM the measurement that, if lost, prevents the construction of the full-rank tree.

The topological method idealized in Costa et al. (1990) allows to identify the CMs through the theory of matroid intersections. In order to determine CSMs, this method removes one redundant measurement from the measurement set at a time, highlighting those that have become critical. These measurements, along with the removed redundant measurement, identify CSMs.

The method proposed in Crainic et al. (1990) is based on the fact that matrices M (incidence matrix of measurements for branches) and A (branch incidence matrix for nodes) are unimodular matrices, that is, matrices that have the property of having the determinants of each square submatrix equal to −1, 1, or 0. In addition to allowing the identification of CMs, this method makes it possible to evaluate the redundancy of the measurement sets at higher levels, which the authors call the strength of the measurement sets. Its disadvantage is to require a combinatorial procedure, which makes the method slow mainly for large networks.

A simple and fast topological method for the identification of CMs has been proposed in Bretas et al. (2005). In this method, the type of the measurements (flow and injection) is explored to reduce the search possibilities. In its first stage, it allows analyzing only the available flow measurements, forming the so-called flow islands. From these flow islands, the redundancy of the injection measurements is analyzed in a more direct way. Note that all injection measurements relating buses belonging to a single flow island are previously identified as redundant. However, the method developed in Bretas et al. (2005) does not allow the identification of CSMs.

According to the literature on the subject, the numerical methods are simpler and easier to implement in computer when compared to the topological ones. However, they are subject to numerical errors. From the methods belonging to this second group, numerical methods, we can highlight the ones proposed in Celik and Liu (1995); Korres and Contaxis (1991a,b); and Coutto Filho et al. (2007).

In general, numerical methods for the identification of CMs and CSMs are based on the following properties of the measured residuals[b] (Celik and Liu, 1995; Coutto Filho et al., 2007):

- CMs have zero residual;
- The elements of the main diagonal of both residual covariance matrix and residual sensitivity matrix[c], corresponding to CMs, are equal to zeros.
- Normalized residuals of measurements from the same CSM are equal in module.
- The correlation coefficients of the measurement residuals from the same CSM are unitary. It is worth remembering that the correlation coefficient of the residuals of two measurements any i and j is given by the following equation:

$$\gamma_{ij} = \frac{\Omega_{ij}}{\sqrt{\Omega_{ii}}\sqrt{\Omega_{jj}}} \tag{5.14}$$

where Ω is the covariance matrix of the residuals[d].

In Ayres and Haley (1986), two algorithms for identifying CSMs are presented. The first is based on the topological definition of CSM while the other on the numerical definition. The two algorithms depend on the analysis of the normalized residuals, thus requiring the execution of the entire state estimation process to identify the CSMs.

[b] It is important to remember that residuals of the measurements are calculated after obtaining the estimates of the state variables.

[c] The residual sensitivity matrix is given by eq. (6.7) presented in Chapter 6.

[d] The covariance matrix of the residuals is given by eq. (6.10) presented in Chapter 6.

Based on the reduced network model developed for the analysis of observability in Contaxis and Korres (1988) and Korres and Contaxis (1991a), an algorithm for the processing of GEs was developed, based on the concept of error residual spread area[e]. They did so considering that, by determining the scattering areas of the residual, it would be possible to reduce the effort that is spent in the processes of detection and identification of measurements with GEs. This is because, once the scattering areas of the residual have been determined, it would be possible to divide the network according to these areas and perform the detection and identification of GEs in each area separately. In order to determine the scattering areas of the residual, it is necessary to perform a search for branches incident only to redundant measurements, thus requiring CMs to be known. The algorithm proposed in Korres and Contaxis (1991a) allows the identification of these measurements by analyzing the elements of the main diagonal of the residual sensitivity matrix.

Using the reduced network model proposed in Contaxis and Korres (1988), as well as graph theory, in Korres and Contaxis (1991b), a method was proposed to identify CMs and CSMs, also allowing the update of these sets when some measurement becomes unavailable. The algorithm reduces the number of calculations required, based on the properties of the so-called flow islands. However, it still needs analysis of the residuals of the measurements, because it makes use of the property of CM to present zero residual and that of the measurements from the same CSM have normalized residuals equal in module.

In Falcao and Arias (1994), a numerical method was developed to allow the analysis of observability and identification of CMs, in a very direct way, through the row-echelon reduction of the Jacobian matrix. The singular aspect of this technique with respect to numerical methods mentioned previously for the identification of CMs is that it does not require the analysis of the properties of the residuals, allowing the identification of CMs in a more direct way.

In Coutto Filho et al. (2007), a method is described for the identification of CMs and CSMs, which makes use of the properties of the measurement

[e] Error residual spread area, or maximum noncritical area, is a set of branches and nodes of the network that are connected only through redundant measurements.

residuals. However, to make the analysis of these properties simpler, the method makes some simplifications because the qualitative characteristics of measurement sets do not depend on the weights and values of the measurements, but rather on the type, location, and number of measurements available. The method requires only a symbolic state estimation, assuming the value 1 both for weighting and for the value of all available measurements.

The method for the identification of CMs and CSMs proposed in Almeida et al. (2008) is based on the triangular factorization of the Gram matrix, which is obtained from the product of the Jacobian matrix of the WLS estimator by the transpose of that matrix. In Almeida et al. (2009), this method was extended to three-phase modeling of the WLS estimator.

Given what has been exposed in this section, the idea behind the development of methods to identify CMs and CSMs, as well as methods for observability analysis, is to obtain procedures that are easy to implement in a computer, requiring a small number of numerical calculations, thus allowing the development of fast computational applications. In order to identify CMs and CSMs, the method developed in London et al. (2007) does not require the analysis of the statistical properties of the measurement residuals, not even any matrix product, but only obtaining and analyzing the structure of the so-called H_Δ matrix, which results from a triangular factorization process of the Jacobian matrix of the WLS estimator. In this way, it allows the identification of CMs and CSMs in a very simple and direct way.

It is important to note, however, that the main motivation for proposing the H_Δ matrix was to allow the possibility of obtaining a procedure that is both simple and easy-to-implement in the computer to identify the level of local redundancy of measurements for state estimation. For this model, a method based on the analysis of the matrix structure H_Δ was presented in London et al. (2007). However, as presented in London et al. (2007), to obtain the matrix H_Δ the EPS must be completely observable. That is, the corresponding Jacobian matrix must have rank equal to the number of state variables to be estimated.

To overcome the problem of obtaining the H_Δ matrix, a method for observability analysis was developed in Benedito et al. (2009). It is based on the process of obtaining the H_Δ matrix, that is, based on the triangular factorization of the Jacobian matrix, as presented in the previous section.

It is important to note that, for observability analysis, only the elimination of the elements of the upper triangle of the Jacobian matrix is required. However, following the LU decomposition procedure, the H_Δ matrix is obtained, making it possible to identify the local redundancy level of the measurements, as well as the observability analysis. It is worth noting that, for systems that are not observable as a whole, but are formed by observable islands, the method proposed in Benedito et al. (2009) makes it possible to identify these observable islands through factorization path trees, also indicating the local redundancy of the measurements of each island, by obtaining and analyzing the H_Δ matrix.

In the next section will be presented the method to identify the level of redundancy through analysis of the H_Δ matrix and then the method to identify CMs and CSMs by analyzing the structure of this matrix.

Note 5.7: As performed to present methods for observability analysis, the methods for analysis of redundancy of measurements based on H_Δ matrix analysis will be presented considering the Jacobian matrix of the linear WLS (H_{DC} matrix) state estimator. It should be noted, however, that they can be applied considering the Jacobian matrix of the nonlinear WLS state estimator, or even the Jacobian matrix of the three-phase version of this estimator, as presented in Toyoshima et al. (2012).

5.2.3 Identification of the local redundancy level through the H_Δ matrix

As defined previously, an EPS will be algebraically observable if the Jacobian matrix H_{DC}, corresponding to the association of this system with a set of measurements, has set equal to the number of state variables to be estimated. It is worth remembering that the H_{DC} matrix relates the active power measurements to the EPS state variables (voltage phase angles). Thus the rows and columns of this matrix correspond, respectively, to the active power measurements and state variables of the EPS.

Given the above, it is possible to conclude that the CMs correspond to the linearly independent rows of that matrix and, consequently, the redundancy ratios place of measurements can be identified by analyzing the linear dependence relations of the H_{DC} matrix rows. However, such relations are difficult to analyze through the matrix structure H_{DC}, but, through a convenient change of base in the space of state variables, that analysis becomes much simplified. Thus in London et al. (2007), the following theorem was demonstrated:

Theorem 5.1 Consider the Jacobian matrix H_{DC} associated with an EPS with n buses and m active power measurements, where $m > (n-1)$. If the system is algebraically observable ($rank(H_{DC}) = n - 1$), then there is a change of base C in the space of the state variables, where in this new base the operator H_{DC} will have the following form (Fig. 5.9):

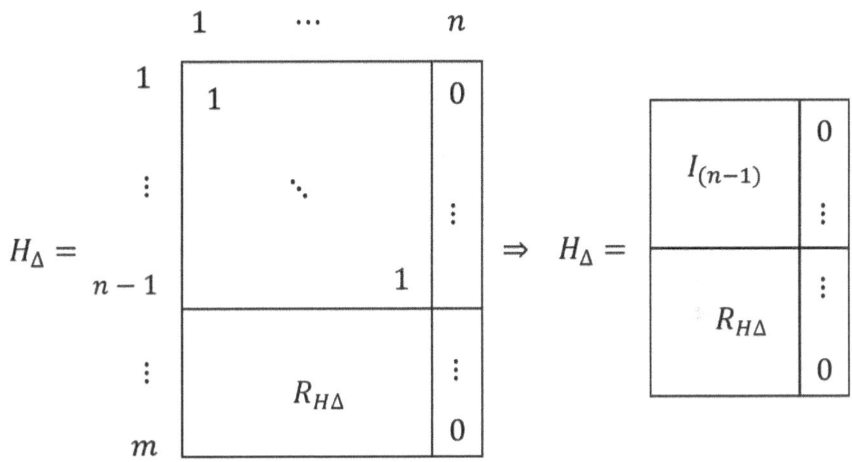

Fig. 5.9 Matrix structure of H_Δ associated with an observable power system with m measurements [$m > (n - 1)$].

where H_Δ is the operator H_{DC} in the new base; I_{n-1} is the identity matrix of dimension $(n - 1) \times (n - 1)$; and $R_{H\Delta}$ is the submatrix of dimension $[m - (n - 1)] \times (n - 1)$, composed by rows linearly dependent on the rows of submatrix I.

Note 5.8: The last column of the H_Δ matrix consists only of zeros because it corresponds to the bus chosen as the angular reference.

In the demonstration of Theorem 5.1, presented in London et al. (2007), it is verified that the matrix H_Δ can be obtained through a triangular factorization process of the H_{DC} matrix, performed by combining the H_{DC} columns. In this way, the matrix H_Δ relates the active power measurements to equivalent state variables, which are linear combinations of the system state variables.

By analyzing the structure of submatrix I, from H_Δ, it is verified that its $(n - 1)$ rows are, separately, linearly independent. Because of this, the measurements corresponding to these rows are called basic measurements, as these measurements are enough to make the system observable. The other measurements are called supplementary.

By analyzing the structure of the H_Δ matrix, we identify the so-called p-critical sets of measurements, through which the local redundancy level of the measurements is identified.

P-critical set of measurements

P-critical set of measurements is a set of p measurements ($p \geq 1$) associated with an observable EPS (or an observable island), which, if lost, make such system unobservable (or unobservable island). However, the removal of any set of k measurements in a p-critical set with ($k < p$) does not cause the loss of system (or island) observability.

Note that a measurement can belong to several p-critical sets, but the p-critical set with the least number of measurements to which it belongs will define its local redundancy level. The definition of local redundancy level based on p-critical sets of measurements is given as follows.

Local redundancy level

A measurement has local redundancy level $p-1$, if the p-critical set, with fewer measurements to which it belongs, has p measurements.

Considering that a CM corresponds to a p-critical set with $p=1$, according to the above definition, it is concluded that the CM has a local redundancy level of 0, as previously presented. If a measurement belongs to at least one critical pair, i.e., p-critical set with $p=2$, it will have a redundancy level of 1 (one), and so on. Considering the structure of the H_Δ matrix, the following lemmas were formulated (London et al., 2007):

Corollary 1: Every supplementary measurement has redundancy level greater than 0.

Lemma 1: Every CM belongs to the set of basic measurements.

Lemma 2: Every p-critical set of measurements has at least one basic measurement.

The search for the p-critical sets of measurements is divided into two phases:

1. Identification of p-critical sets of measurements containing only one basic measurement.
2. Identification of p-critical sets of measurements containing more than one basic measurement.

Next, it will be shown that the second phase of the search is a recursive application of the first. Consider, then, Theorem 5.2, whose proof is presented in London et al. (2007).

Theorem 5.2 The p measurements corresponding to the rows of p nonzero elements that belong to the same column of the H_Δ matrix form a p-critical set of measurements containing only one basic measurement.

Theorem 5.2 shows that when a column has only one nonzero element, it means that the equivalent state information corresponding to that column is provided by only one measurement, so this measurement is critical (it has local redundancy level equal to 0).

In order to carry out the second phase of the search using the guidelines of Theorem 5.2, a noncritical basic measurement of the H_Δ matrix is eliminated to obtain the new H_Δ matrix. Since the removed basic measurement corresponds to a row linearly dependent on at least one row of the submatrix R, there is at least one supplementary measurement that can replace it. By performing the substitution, we obtain the new matrix H_Δ. Analyzing the columns of this new matrix, considering Theorem 5.2, we conclude that the p measurements associated with the new p-critical sets identified constitute, together with the basic measurement that was removed, a $(p+1)$-critical set of measurements containing two basic measurements.

Applying this process to all basic measurements with a redundancy level greater than or equal to 1 (i.e., noncritical measurements), all p-critical sets containing two basic measurements are identified. By continuing this process, by eliminating b sets of basic measurements, one in each time, with local redundancy level greater than or equal to b, all p-critical sets of measurements, with $p \geq (b+1)$, containing $(b+1)$ basic measurements will be identified.

From the analysis of the p-critical sets of measurements, the local redundancy level of all available measurements is identified.

According to the reference (Crainic et al., 1990), in terms of local redundancy of measurements, a measurement system consisting of measurements with a redundancy level >2 is considered reliable. This is because, for such a measurement system, the absence of CMs (with local redundancy level equal to 0) and CSMs would be guaranteed (measurements that constitute a critical set of measurements have a local redundancy level of 1). Thus the search for

p-critical sets of measurements for, at most $p=3$, allowed the identification of measurements with redundancy level 0, 1, and 2 (London et al., 2007).

Note 5.9: From this point on, the method to identify the local redundancy level will be presented considering the factorization of the transposed Jacobian matrix H_{DC}^T, which will then originate the transpose of the matrix H_Δ. As mentioned previously in the presentation of the method for analysis of observability based on the triangular factorization of the H_{DC} matrix, if we consider the structure of the H_{DC}^T matrix, it becomes easier to apply the factorization process in this matrix instead of applying directly to the H_{DC} matrix. The application of the triangular factorization process in the H_{DC}^T matrix associated with an observable EPS will then result in the H_Δ^T matrix, the structure of which is shown in Fig. 5.10.

$$H_\Delta^T = \begin{array}{c} \\ 1 \\ \vdots \\ n-1 \\ n \end{array}\begin{array}{c} 1 \quad \cdots \quad n-1 \quad \cdots \quad m \\ \left[\begin{array}{ccc|c} 1 & & & \\ & \ddots & & R_{H\Delta}^T \\ & & 1 & \\ \hline 0 & \cdots & & \cdots \quad 0 \end{array}\right] \end{array} \Rightarrow H_\Delta^T = \left[\begin{array}{c|c} I_{(n-1)} & R_{H\Delta}^T \\ \hline 0 \quad \cdots & \cdots \quad 0 \end{array}\right]$$

Fig. 5.10 Matrix structure H_Δ^T associated with an observable power system with m measurements $[m>(n-1)]$.

where H_Δ^T: H^T is the operator on new base; I_{n-1} is the identity matrix of dimension $(n-1)\times(n-1)$; and $R_{H\Delta}^T$ is the dimension submatrix $(n-1)\times[m-(n-1)]$, composed of columns linearly dependent on columns of submatrix I_{n-1}.

Method description

The method for identifying the local redundancy level proposed in London et al. (2007) requires making calculations. Given that the level of redundancy of the measurements depends only on the type and location of the available measurements, but not the network parameters values, in order to minimize the probability of rounding errors, the H_{DC}^T matrix is constructed by assigning the value 1 to line reactances.

The method consists of four stages, which are as follows:

Stage 1: Obtaining and performing triangular factorization of the H_{DC}^T matrix.

Stage 2: Identification of measurements with redundancy level 0 and critical pairs and triplets of measurements involving only one basic measurement.

Stage 3: Identification of critical pairs and triplets of measurements involving more than one basic measurement.

Stage 4: Identification of measurements with local redundancy levels 1 and 2.

Then each of these stages will be described, considering the application of the Gauss elimination process in the H_{DC}^T matrix.

First stage

The Gauss elimination process, used in the method, has some peculiarities, even applied to the H_{DC}^T matrix. Before introducing them, the steps of the Gauss elimination process will be presented, which are as follows:

1. FORWARD: The nonzero elements of the lower triangle of the matrix under analysis are made zeros by linear combinations of the rows of this matrix. The result of this stage is an upper triangular matrix.
2. DIAGONAL OPERATION: We make equal to 1 all elements of the main diagonal of the upper triangular matrix obtained in the previous step by dividing each row of this matrix by the corresponding element of its main diagonal.
3. BACKWARD: The nonzero elements of the upper triangle of the matrix resulting from the DIAGONAL OPERATION are made zeros by linear combinations of the rows of this matrix. The result of this stage is an identity matrix.

As mentioned in the previous section, the definition of algebraic observability, applied to the H_{DC} matrix, also applies to the H_{DC}^T matrix. Thus as the necessary condition for the system to be algebraically observable is that the rank of the H_{DC}^T matrix is equal to $(n-1)$, the FORWARD process will be performed up to the $n-1$ diagonal. To perform the FORWARD step, the method adopts the columns deletion scheme. After performing this step, the last row of the matrix under factorization will consist only of zeros, because this row will correspond to the bus chosen as the angular reference.

The step "DIAGONAL OPERATION" will then be performed considering the $(n-1)$ H_{DC}^T matrix rows; then the BACKWARD step is performed, through which the nonzero elements of the upper part of the H_{DC}^T submatrix resulting from the previous step of dimension $(n-1) \times (n-1)$ are eliminated (makes equal to 0). It is necessary, however, that the BACKWARD step is performed through linear combinations to be made with the rows of that matrix, since in this way all columns of the H_{DC}^T matrix will be considered.

The H_{Δ}^T matrix will be obtained after performing the BACKWARD step.

Second stage

We will identify the measurements with redundancy 0 (CMs) and the critical pairs and triplets of measurements involving only one of the basic measurements. This will be done through a search for nonzero elements in the rows of the matrix H_{Δ}^{T} (according to Theorem 5.2).

Third stage

Through new triangular factorizations of the H_{Δ}^{T} matrix, as described above, the critical pairs and triplets of measurements containing more than one basic measurement will be identified.

Fourth stage

The measurements with local redundancy levels 1 and 2 will be identified through the critical pairs and triplets of measurements, as follows:

1. Measurement with local redundancy level 1: It is the noncritical measurement that appears in at least one critical pair of measurements;
2. Measurement with local redundancy level 2: It is the noncritical measurement that does not appear in any critical pair of measurements, but in at least one critical triplet of measurements.

Algorithm for application in observable EPS

Before presenting the steps of the algorithm, it is important to highlight that the H_{cc}^{T} matrix will be formed considering, first, the flow measurements due to the fact that the injection measurements can relate more than two variables, which, as a consequence, would imply more than two nonzero elements in the H_{cc}^{T} matrix columns associated with these measurements. This is done to decrease the number of elements generated during the FORWARD process (these elements are known as fill-ins). Finally, the steps of the algorithm are as follows:

- **Step 1:** With the available measurement set, obtain the matrix H_{Δ}^{T}.
- **Step 2:** Perform FORWARD processing to the pivot $(n-1)$.
- **Step 3:** Carry out the "DIAGONAL OPERATION".
- **Step 4:** Perform BACKWARD processing.
- **Step 5:** Check, in the rows of the resulting H_{Δ}^{T} matrix, the number of nonzero elements that appear. From these elements, identify the measurements with local redundancy level 0 (CMs), as well as the critical pairs and triplets of measurements involving only one of the basic measurements.
- **Step 6:** Through new triangular factorizations of the H_{Δ}^{T} matrix, identify the critical pairs and triplets of measurements involving more than one of the basic measurements.

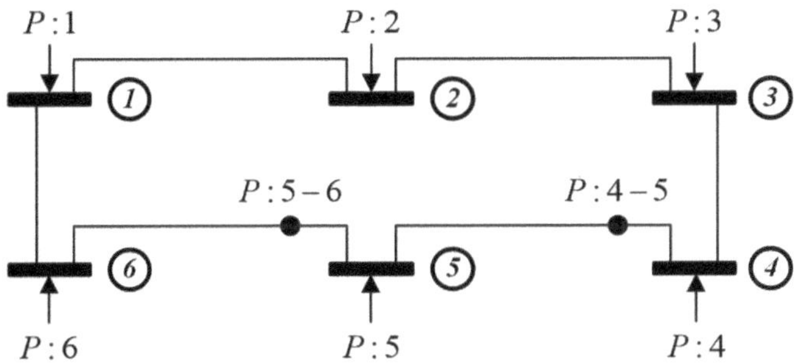

Fig. 5.11 System of six buses associated with an active measurement set (•, power flow measurements; →, power injection measurements).

- **Step 7:** Identify the measurements with redundancy levels 1 and 2 through the critical pairs and triplets of measurements identified in the previous steps.

Note 5.10: When performing the FORWARD step, if a null pivot is found, look for another column from the pivot column that has a nonzero element in the pivot row. Once this column is found, replace it with the null pivot column. Thus we will have a new nonnull pivot, and the FORWARD step can continue[f].

Example 5.2a: To illustrate the application of the algorithm mentioned above, the six-bus EPS associated with the measurement set shown in Fig. 5.11 will be used.

Step 1: Construct the transposed Jacobian matrix H^T considering, first, the flow measurements.

$H^T =$

	P:4-5	P:5-6	P:1	P:2	P:3	P:4	P:5	P:6
1	0	0	2	-1	0	0	0	-1
2	0	0	-1	2	-1	0	0	0
3	0	0	0	-1	2	-1	0	0
4	1	0	0	0	-1	2	-1	0
5	-1	1	0	0	0	-1	2	-1
6	0	-1	-1	0	0	0	-1	2

After Gauss elimination (Steps 2–4), the following H_Δ^T matrix is obtained:

[f] If the system under analysis is observable, it will always be possible to find a column with a nonzero element in the null pivot row.

$H_{\Delta}^{T} =$

	P:1	P:2	P:3	P:4-5	P:5-6	P:4	P:5	P:6
1	1	0	0	0	0	-0.25	0	-0.75
2	0	1	0	0	0	-0.50	0	-0.50
3	0	0	1	0	0	-0.75	0	-0.25
4	0	0	0	1	0	1.25	-1	-0.25
5	0	0	0	0	1	0.25	1	-1.25
6	0	0	0	0	0	0	0	0

Step 5: By analyzing the rows of the obtained H_{Δ}^{T} matrix, we identify:

- First row: a critical triplet formed by measurements [P:1; P:4; P:6];
- Second row: a critical triplet formed by measurements [P:2; P:4; P:6];
- Third row: a critical triplet formed by measurements [P:3; P:4; P:6];
- Fourth row: has more than three nonzero elements, thus no critical pair or triplet of measurements are identified;
- Fifth row: it also has more than three nonzero elements.

Step 6: Considering: $\begin{cases} n=6 \\ m=8 \end{cases} \rightarrow m-(n-1)=8-(6-1)=3.$

Since $m-(n-1)=3$, it will be necessary to identify critical pairs and triplets of measurements involving more than one of the basic measurements. If $m-(n-1)\leq 1$, it would not be necessary to search for pairs and triplets of critical measurements involving more than one of the basic measurements since any two measurements that were simultaneously lost would render the system unobservable.

(i) Removing from the matrix H_{Δ}^{T} the first column, which corresponds to the measurement P:1, we obtain:

$H^{T}(P{:}1) =$

	P:2	P:3	P:4-5	P:5-6	P:4	P:5	P:6
1	0	0	0	0	-0.25	0	-0,75
2	1	0	0	0	-0.50	0	-0,50
3	0	1	0	0	-0.75	0	-0,25
4	0	0	1	0	1.25	-1	-0,25
5	0	0	0	1	0.25	1	-1,25
6	0	0	0	0	0	0	0

Changing the column corresponding to the measurement P:4, the matrix H^T (P:1) becomes:

$H^T(P{:}1) =$

	P:4	P:2	P:3	P:4-5	P:5-6	P:5	P:6
1	-0.25	0	0	0	0	0	-0.75
2	-0.50	1	0	0	0	0	-0.50
3	-0.75	0	1	0	0	0	-0.25
4	1.25	0	0	1	0	-1	-0.25
5	0.25	0	0	0	1	1	-1.25
6	0	0	0	0	0	0	0

From the matrix above, we obtain the following:

$H^T{}_{\Delta(P{:}1)} =$

	P:4	P:2	P:3	P:4-5	P:5-6	P:5	P:6
1	1	0	0	0	0	0	3
2	0	1	0	0	0	0	1
3	0	0	1	0	0	0	2
4	0	0	0	1	0	-1	-4
5	0	0	0	0	1	1	-2
6	0	0	0	0	0	0	0

Through this matrix, we identify:

- First row: a critical triplet formed by measurements [P:1; P:4; P:6], which is a critical triplet involving only one of the basic measurements, the P:1 measurement, which had already been identified through the H_Δ^T matrix;
- Second row: a critical triplet formed by measurements [P:1; P:2; P:6];
- Third row: a critical triplet formed by measurements [P:1; P:3; P:6];
- Fourth row: has more than two nonzero elements, thus not identifying any critical pair or triplet of measurements;
- Fifth row: it also has more than two nonzero elements.

For the example above, it will be necessary to identify all critical triplets of measurements including those formed only by basic measurements. Thus starting from the matrix $H_{\Delta(P:1)}^T$, the columns corresponding to the basic measurements associated with the $(n-1)$ columns of the H_Δ^T matrix will be removed from this matrix one at a time.

(ii) Removing the second column from the matrix $H_{\Delta(P:1)}^T$, which corresponds to the measurement P:2, one obtains the following matrix:

$H^T(P{:}1, P{:}2) =$

	P:4	P:3	P:4-5	P:5-6	P:5	P:6
1	1	0	0	0	0	3
2	0	0	0	0	0	1
3	0	1	0	0	0	2
4	0	0	1	0	-1	-4
5	0	0	0	1	1	-2
6	0	0	0	0	0	0

Changing the column corresponding to the measurement P:6 and performing the operations already known, the matrix $H^T_{(P:1,P:2)}$ becomes:

$H^T\Delta(P{:}1, P{:}2) =$

	P:4	P:6	P:3	P:4-5	P:5-6	P:5
1	1	0	0	0	0	0
2	0	1	0	0	0	0
3	0	0	1	0	0	0
4	0	0	0	1	0	-1
5	0	0	0	0	1	1
6	0	0	0	0	0	0

Through the matrix $H^T_{\Delta(P:1,P:2)}$, we identify:

- First row: a critical triplet formed by measurements [P:1; P:2; P:4];
- Second row: a critical triplet formed by measurements [P:1; P:2; P:6], which had already been identified through the $H^T_{\Delta(P:1)}$ matrix;
- Third row: a critical triplet formed by measurements [P:1; P:2; P:3];
- Fourth row: it has more than one nonzero element; therefore it does not identify any critical triplet of measurements;
- Fifth row: analogous to the previous row.

(iii) Removing the third column from the matrix $H^T_{\Delta(\text{P}:1)}$, which corresponds to the measurement P:3, one obtains the following matrix:

$H^T(P{:}1, P{:}3) =$

	P:4	P:2	P:4-5	P:5-6	P:5	P:6
1	1	0	0	0	0	3
2	0	1	0	0	0	1
3	0	0	0	0	0	2
4	0	0	1	0	-1	-4
5	0	0	0	1	1	-0.50
6	0	0	0	0	0	0

Through the operations already presented, we obtain the following matrix:

$H^T{}_{\Delta(P:1,\, P:3)} =$

	P:4	P:2	P:6	P:4-5	P:5-6	P:5
1	1	0	0	0	0	0
2	0	1	0	0	0	0
3	0	0	1	0	0	0
4	0	0	0	1	0	-1
5	0	0	0	0	1	1
6	0	0	0	0	0	0

Not considering the critical triplet already identified, it is now identified only the critical triplet formed by measurements [P:1; P:3; P:4].

(iv) Removing the fourth column from the matrix $H^T_{\Delta(P:1)}$ corresponding to the measurement P:4–5, results in the following matrix:

$H^T{}_{(P:1,\, P:4\text{-}5)} =$

	P:4	P:2	P:3	P:5-6	P:5	P:6
1	1	0	0	0	0	3
2	0	1	0	0	0	1
3	0	0	1	0	0	2
4	0	0	0	0	-1	-4
5	0	0	0	1	1	-0.50
6	0	0	0	0	0	0

From the previous matrix, we obtain:

$H^T{}_{\Delta(P:1,\, P:4\text{-}5)} =$

	P:4	P:2	P:3	P:5	P:5-6	P:6
1	1	0	0	0	0	3
2	0	1	0	0	0	1
3	0	0	1	0	0	2
4	0	0	0	1	0	-4
5	0	0	0	0	1	-0.50
6	0	0	0	0	0	0

Since each row in the matrix above has more than one nonzero element, no critical triplet of measurements can be identified.

(v) Removing the fifth column from the matrix $H^T_{\Delta(P:1)}$, which corresponds to the measurement P:5–6, we obtain the following matrix:

$H^T(P{:}1) =$

	P:4	P:2	P:3	P:4-5	P:5-6	P:5	P:6
1	-0.25	0	0	0	0	0	-0.75
2	-0.50	1	0	0	0	0	-0.50
3	-0.75	0	1	0	0	0	-0.25
4	1.25	0	0	1	0	-1	-0.25
5	0.25	0	0	0	1	1	-1.25
6	0	0	0	0	0	0	0

where it is obtained:

$H^T_{\Delta}(P{:}1) =$

	P:4	P:2	P:3	P:4-5	P:5-6	P:5	P:6
1	1	0	0	0	0	0	3
2	0	1	0	0	0	0	1
3	0	0	1	0	0	0	2
4	0	0	0	1	0	-1	-4
5	0	0	0	0	1	1	-2
6	0	0	0	0	0	0	0

Analogously to the previous case, it is not possible to identify any critical triplet of measurements.

Continuing the process for all columns and all sets of two columns of the H^T_{Δ} matrix corresponding to the basic measurements, the following critical pairs and triplets of measurements involving more than one basic measurement are identified:

[P:1; P:2; P:6], [P:1; P:3; P:6], [P:1; P:2; P:4], [P:1; P:2; P:3], [P:1; P:3; P:4]; [P:2; P:3; P:6], and [P:2; P:3; P:4].

Step 7: Through the information obtained in Steps 5 and 6, we arrive at the following result:

- P:1 = local redundancy level 2;
- P:2 = local redundancy level 2;
- P:3 = local redundancy level 2;
- P:4–5 = local redundancy level greater than 2, since this redundant measurement did not appear in any critical pair and triplet of measurements;

- P:5–6 = local redundancy level greater than 2, since this redundant measurement also did not appear in any critical triplet of measurements;
- P:4 = local redundancy level 2;
- P:5 = local redundancy level greater than 2;
- P:6 = local redundancy level 2.

5.2.4 Identification of CMs and CSMs from the H_Δ^T matrix analysis

As shown in London et al. (2007), to identify the CMs from the structure of the matrix H_Δ, it is sufficient to carry out a search for the H_Δ^T rows that have only one nonzero element. The measurements that correspond to these columns are critical.

To identify CSMs from the structure of the H_Δ^T matrix, we initially identify the p-critical sets of measurements with $p = 2$, i.e., the critical pairs of measurements. This is because the two measurements that constitute a critical pair belong to the same CSM (London et al., 2004).

It is important to clarify that a p-critical set is not, by definition, the same as a CSM because, according to its topological definition presented earlier in this chapter, CSM is the set of measurements constituted by redundant measurements in which the elimination of one of its measurements renders the other measurements critical. On the other hand, a p-critical set of measurements will be equal to a CSM only when both have only two measurements, since the removal of one of the measurements from one critical pair makes the other measurement critical. However, p-critical sets with $p \neq 2$ do not constitute CSMs. For example, the p-critical set with $p = 1$ is a CM itself. In contrast, in a p-critical set with $p = 3$, it is verified that the excluding of one of its measurements does not make the other measurements critical.

Considering what was discussed above, a method for identification of CMs and CSMs was presented in London et al. (2004), based on the H_Δ^T matrix structure analysis, which consists of the following steps:

- **Step 1**: Using the available measurements, derive the H_{cc}^T matrix, thus obtaining the matrix H_Δ^T. Then identify, through the rows of this matrix, CMs and critical pairs of measurements formed by only one basic measurement.
- **Step 2**: Among the critical pairs identified in Step 1, select those that do not have a common supplementary measurement. The two measurements of each of these critical pairs constitute a CSM containing only two measurements.

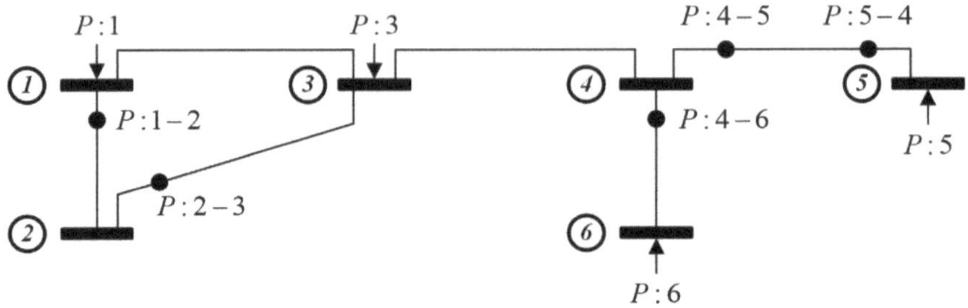

Fig. 5.12 Six-bus test system associated with an active measurement set (•, power flow measurements; →, power injection measurements).

- **Step 3**: Among the critical pairs identified in the Step 1, select groups that have at least one common supplementary measurement. The measurements pertaining to each of these groups constitute a single CSM containing more than two measurements.
- **Step 4**: If there are any noncritical basic measurements not belonging to the CSMs already identified, the measurements (i.e., the column corresponding to this measurements) are eliminated from the H_Δ^T matrix. Then, the new H_Δ^T matrix is obtained, and analyzing the rows of this matrix, the basic measurements, which are now identified as critical, will constitute, along with the basic measurements eliminated, a CSM.

Step 4 is finalized when all noncritical basic measurements not belonging to CSMs already identified have been analyzed. In the worst case in terms of processing, it would be necessary a maximum of $(n-1)$ partial rechecks, accompanied by counts of nonzero elements in $(n-1)$ matrices (London et al. 2004). Such a situation would occur when p-critical set with $p \leq 2$ was identified in the first step.

The numerical method proposed in London et al. (2004) is briefly described in this section, based on the analysis of the structure of the H_Δ^T matrix and does not require the residual analysis, nor obtaining and analysis of the elements of the matrices covariance and sensitivity of the residual. In this way, it is a method of easy implementation in computer that does not require great amount of numerical calculation. It should be noted that, by using sparsity techniques and the linear dependence relationships of the H_Δ^T matrix rows, it is possible to increase the computational efficiency of this method for the identification of CMs and CSMs. More specifically, it is possible to minimize the number of partial refactorizations required in Step 4 of the process.

Example 5.2b: To illustrate the application of the method for identifying CMs and CSMs through the H_Δ^T matrix, the six-bus EPS associated with the set of measurements shown in Fig. 5.12 will be used.

Step 1: Considering the measurements shown in Fig. 5.12, the following H^T matrix is obtained (the line reactances are assigned the value 1 pu):

$H^T(P{:}1, P{:}2) =$

	P:4	P:3	P:4-5	P:5-6	P:5	P:6
1	1	0	0	0	0	3
2	0	0	0	0	0	1
3	0	1	0	0	0	2
4	0	0	1	0	-1	-4
5	0	0	0	1	1	-2
6	0	0	0	0	0	0

The matrix H_Δ^T will be

$H^T{}_{\Delta(P{:}1, P{:}2)} =$

	P:4	P:6	P:3	P:4-5	P:5-6	P:5
1	1	0	0	0	0	0
2	0	1	0	0	0	0
3	0	0	1	0	0	0
4	0	0	0	1	0	-1
5	0	0	0	0	1	1
6	0	0	0	0	0	0

Analyzing the rows of this matrix, a critical measurement, P:3, is identified, as well as the following three critical pairs of measurements:

- [P:1–2, P:1];
- [P:2–3, P:1];
- [P:4–6, P:6].

Step 2: The critical pair [P:4–6, P:6] is the only one related with the supplementary measurement P:6. Therefore it constitutes a CSM formed only by a basic measurement.

Step 3: Supplementary measurement P:1 is present in two critical pairs: [P:1–2, P:1] and [P:2–3, P:1]. Consequently, the measurements belonging to these critical pairs constitute a single critical set of measurements formed by three measurements: [P:1–2, P:2–3, P:1].

Step 4: P:5–4 is the only noncritical basic measurement, which did not appear in any critical set identified in the previous step. Thus by eliminating this measurement of the H_Δ^T matrix, we obtain:

$$H^T_{(P:1,\,P:3)} = \begin{array}{c} \\ 1 \\ 2 \\ 3 \\ 4 \\ 5 \\ 6 \end{array} \left[\begin{array}{cccc|cc} P{:}4 & P{:}2 & P{:}4\text{-}5 & P{:}5\text{-}6 & P{:}5 & P{:}6 \\ 1 & 0 & 0 & 0 & 0 & 3 \\ 0 & 1 & 0 & 0 & 0 & 1 \\ 0 & 0 & 0 & 0 & 0 & 2 \\ 0 & 0 & 1 & 0 & -1 & -4 \\ 0 & 0 & 0 & 1 & 1 & -0.50 \\ \hline 0 & 0 & 0 & 0 & 0 & 0 \end{array}\right]$$

By permutating the columns of this matrix, as shown below, we can obtain the new matrix H_Δ^T, shown below:

$$H^T_{\Delta(P:1,\,P:3)} = \begin{array}{c} \\ 1 \\ 2 \\ 3 \\ 4 \\ 5 \\ 6 \end{array} \left[\begin{array}{ccccc|c} P{:}4 & P{:}2 & P{:}6 & P{:}4\text{-}5 & P{:}5\text{-}6 & P{:}5 \\ 1 & 0 & 0 & 0 & 0 & 0 \\ 0 & 1 & 0 & 0 & 0 & 0 \\ 0 & 0 & 1 & 0 & 0 & 0 \\ 0 & 0 & 0 & 1 & 0 & -1 \\ 0 & 0 & 0 & 0 & 1 & 1 \\ \hline 0 & 0 & 0 & 0 & 0 & 0 \end{array}\right]$$

By analyzing the rows of this matrix, we note that no basic measurement has become critical. Therefore P:5–4 is not part of any critical set and the analysis ends.

The following result is obtained:

- Critical measurement: P:3;
- Critical sets: [P:1, P:1–2, P:2–3] and [P:4–6, P:6].

5.3 Qualitative characteristics for metering systems containing SPMs and SCADA measurement

This section will be developed based on the WLS hybrid estimator presented in Chapter 4, which enables the simultaneous processing of SCADA and SPMs. All the concepts and definitions already presented in this chapter considering metering systems formed only by SCADA measurements are valid for hybrid systems, i.e., systems containing SCADA measurements and SPMs, provided that in the presence of at least a PMU voltage-phase angle, the following condition for algebraic observability for a SEP with n buses is met:

$$Rank(H_{Hib}) = 2n. \tag{5.15}$$

This happens because no bus will be chosen as the angular reference since the voltage phase angle provided by PMUs is obtained from another reference frame, which is determined by the instant of time provided by the reception of the signal emitted by the GPS system.

It is worth noting, however, that to consider observability and redundancy of measurements for application of the WLS hybrid estimator through the $P\theta$ model (also known as linear model), some special considerations are necessary. This is because, in the presence of current SPMs, the decoupling $P\theta$ and QV from the observability problem are not immediate. To verify this particularity, it is enough to analyze the current phasor derivatives in the Jacobian matrix of the WLS hybrid state estimator presented in Zhu and Abur (2007). Thus the methods for analyzing the qualitative characteristics of sets of measurements based on the triangular factorization of the Gain and Jacobian matrices presented in this chapter are usually applied considering the complete matrices of hybrid systems, that is, without decoupling $P\theta$ and QV.

Just to make it clearer, a system having SCADA measurements and SPMs will be declared observable if during the triangular factorization of the corresponding Gain matrix (or Jacobian matrix) no null pivot is found.

Considering the above, to allow the application of the $P\theta$ model for observability and redundancy analysis of measurements considering the WLS hybrid estimator, some additional considerations and/or approximations are necessary in order to make it viable. It will be considered a PMU having a number of communication channels in order to be possible measuring the voltage phasor at bus i, where it is installed, as well as the current phasor from bus i to bus j. Also considering that the impedance of all branches is equal to j1 pu and the magnitudes of voltages in all the buses are equal to 1.0 pu, the real part of the current phasor from bus i to bus j (I_{ij}) can be modeled as (Chen and Abur, 2006):

$$real(I_{ij}) \cong \theta_i - \theta_j, \tag{5.16}$$

where θ_i and θ_j are the voltage phase angles of the buses i and j, respectively. Thus the rows of the Jacobian matrix of the linear hybrid WLS estimator (model $P\theta$) corresponding to the SPM of the voltage phase angle in the bus i (m_{θ_i}) and the real part of the SPM of the current of bus i to bus j ($m_{C_{i-j}}$) can be represented as (Chen and Abur, 2006; London et al., 2009):

$$
H_{\theta PMU} = \begin{matrix} \vdots \\ \theta_i \\ I_{ij} \\ \vdots \end{matrix} \overset{\begin{matrix} \quad\quad \theta_i & \theta_j \end{matrix}}{\begin{bmatrix} \cdots & \vdots & \cdots & \vdots & \cdots \\ \cdots & 1 & \cdots & \vdots & \cdots \\ \cdots & 1 & \cdots & -1 & \cdots \\ \cdots & \vdots & \cdots & \vdots & \cdots \end{bmatrix}} \tag{5.17}
$$

Following the steps as described previously, it will be possible to carry out the observability and redundancy analysis of measurements, for application of the linear hybrid WLS estimator, through a process of factorization of the Jacobian matrix of the model $P\theta$, similar to the one performed for measurement systems, but formed only by conventional SCADA measurements, as presented in London et al. (2009).

5.4 Planning and reinforcement of metering systems

In addition to being noise carriers, which is inherent to the measurement process, the analog measurements used in the PSSE process are subject to GEs. It is also worth noting that they may be momentarily inaccessible to the state estimator due to some problem in the telemetry system, thus causing the unavailability of communication channels.

In the literature, this unavailability is usually referred to as loss of measurements and/or remote terminal units (RTUs) (electronic equipment responsible for collecting system measurements at power plants and substations and for sending them to operation centers). The results of the PSSE process depend on the available measurement system, i.e., the number, type, and location of the meters and the installed RTUs (Coser et al., 2006; Vigliassi et al., 2009).

In this sense, it is usually defined as a reliable measurement system, the observable system free of CMs, CSMs, critical RTUs[g], and, for the application of the hybrid WLS estimator, critical PMUs[h] (Souza et al., 2005; Coser et al., 2006; Vigliassi et al., 2019). It is important to clarify the following points:

[g] An RTU is critical if the simultaneous elimination of all of its measurements makes an observable SEP unobservable.

[h] A PMU is critical if the loss of its measurements turns an observable SEP into an unobservable one.

1. As mentioned previously, to be observable means that the available analog measurements make it possible to determine all the EPS state variables, i.e., the complex voltages in all the EPS buses.
2. Absence of CMs and CSMs enables the processing of GEs in analog measurements (as it will be presented in the next chapter). That is, there is no possibility to detect GEs in CMs (Clements et al., 1981), nor even identify such errors in measurements in CSMs (Mili et al., 1984), when the WLS estimator, associated with GE processing techniques based on the analysis of the residuals of the estimation process, is used (as it will be discussed in the next chapter).
3. If observable and without CMs, CSMs, critical RTUs, and critical PMUs, the system will remain observable even with the loss of any one or two measurements, or even loss of all measurements transmitted by any RTU or PMU.

Considering only the performance requirements mentioned above, which are related to the qualitative characteristics of a measurement system, i.e., observability and redundancy of measurements, RTUs and PMUs, the best solution to the problem of metering system planning (MSP) for PSSE would be to install meters, UTRs and PMUs in all possible locations. However, in addition to the qualitative characteristics, another important aspect in the planning of a metering system is the cost associated with the acquisition and installation of meters, RTUs and PMUs, due to the budget constraints of electric power companies.

The planning of metering systems with no cost considerations is already a combinatorial optimization problem with several constraints, which requires the investigation of a large number of possible solutions (London et al., 2003; Milosevic and Begovic, 2003; Coser et al., 2006), being considered a difficult problem to solve (Brueni and Heath 2005). When the cost of acquiring and installing meters, RTUs and PUMs, is considered, the problem becomes even more complicated due to the conflicting requirements between the two objectives: (1) to minimize the cost of acquiring and installing equipment, and (2) maximizing the qualitative characteristics (observability and absence of CMs, CSMs, critical UTRs, and critical PMUs). Therefore due the financial requirements, the MSP is a multiobjective, combinatorial optimization problem, which involves conflicting objectives.

In order to solve this complex problem, methods developed for the analysis of observability and redundancy of measurements have been combined with metaheuristics (Milosevic and Begovic, 2003; Souza et al., 2005; Coser et al., 2006; Müller and Castro, 2016). As, for example, the MSP method proposed in Vigliassi et al. (2019), which combines a multiobjective evolutionary algorithm based on subpopulation tables with the properties of the H_Δ matrix. This method can solve the problem considering the existence of only measurements from SCADA systems, only from PMUs of both types of measurements.

5.5 Update of the qualitative characteristics of measurement sets

Having a reliable metering system is a necessary but not sufficient condition for the success of the PSSE process. This is because, during the process of real-time operation of the EPSs, there is a possibility of failure in the telemetry system, causing the temporary loss (or unavailability) of a number of measurements and/or RTUs (or PMUs) greater than anticipated in the planning of the metering system. In such situations, it is necessary to update the qualitative characteristics of the measurement set available to successful perform the PSSE process, thus making possible a reliable operation of the EPS. To achieve this goal, it is necessary to obtain the following information about loss of measurements (London et al., 2007):

1. Whether the system remains observable; and
2. If not, which pseudo measurements are required to restore the system observability
3. What are the new CMs and new CSMs, if they exist, to keep aware of weak spots in the actual measurement set

Then a method based on the structure of the H_Δ matrix was developed in London et al. (2007), which allows not only the analysis of the qualitative characteristics of the aforementioned measurement set (analysis and restoration of observability and identification of CMs and CSMs), but also a rapid update of these characteristics after loss of measurements and/or RTUs.

In order to allow fast update of the qualitative characteristics of measurement sets following the loss of measurements and/or RTUs, the method proposed in London et al. (2007) performs a "pre-analysis" of the data before it is put into operation, i.e., before analyzing a sample of measurements that becomes available in the Operation Center. The preanalysis of the data is carried out considering that the measurements of all the meters installed in the system are available, as well as all the virtual measurements. This initial situation is called Base Case. The base case analysis consists of:

- Analysis 1: Construction of the Jacobian matrix H of the linear WLS state estimator and then obtaining the matrix H_Δ, storing the triangular factors;
- Analysis 2: Identification of the p-critical sets of measurements, formed by only one basic measurement (through the rows of the H_Δ matrix, obtained in analysis 1);
- Analysis 3: Identification of CMs and CSMs, as presented in Section 5.2.4.

In the moment when a sample of measurements becomes available at the Operation Center, the proposed method performs the following steps:

- **Step 1:** If there is no loss of measurements, that is, if the available measurements are the same as those considered in the base case, no analysis is necessary because the qualitative characteristics did not change.
- **Step 2:** If only supplementary measurements have been lost, the method will indicate that the system remains observable and then checks the "criticality", so to speak, of the measurements available, i.e., it verifies the existence of CMs and CSMs. For this purpose, it is enough to eliminate the rows corresponding to the lost measurements from the matrix H_Δ obtained for the base case. Next, the CMs and CSMs are identified, applying the method presented in Section 5.2.4 to the new H_Δ matrix. It is worth mentioning, however, that in view of the qualitative analyses carried out initially for the base case, it is possible to increase the efficiency of the identification process of CMs and CSMs through the method presented in Section 5.2.4.
- **Step 3:** If at least one basic measurement has been lost, the system may then become unobservable. Thus before verifying the "criticality" of the available measurements, the observability analysis is performed as follows:
  - **(a)** The rows corresponding to the measurements that were lost are eliminated from the H_Δ matrix obtained for the base case. Then, the number of columns formed by only zeros is verified in the matrix resulting from the elimination of the missing measurements. Since basic measurements have been lost, the matrix resulting from the elimination of the lost measurements will no longer be a H_Δ matrix, once it will no longer have the structure shown in Fig. 5.9.
  - **(b)** If the matrix resulting from step (i) has only the last column formed by zeros, it is evidence that the EPS is still observable as a whole and we can proceed to obtaining the new matrix H_Δ by the permutation of columns of the resulting matrix of step (i). Next, CMs and CSMs are identified, using the method presented in Section 5.2.4.

(c) If at least one additional column of zeros appears in the matrix resulting from step (i), it is because the system has lost observability, since there is no measurement giving information of the equivalent state associated with that column. In this situation, we identify the pseudo measurements necessary to restore the observability of the SEP through the triangular factors obtained for the base case as presented in section "Method based on the triangular factorization of the Jacobian matrix and the concept of factorization paths." Then, the new H_Δ matrix is obtained and the CMs and CSMs are identified using the method presented in Section 5.2.4.

It is worth mentioning that all the analyses performed by the method proposed in London et al. (2007) to update the qualitative characteristics of the measurement set mentioned above require only partial re-evaluations of the H_Δ matrix obtained for the base case. In such way, by using sparsity techniques, the method provides an algorithm that is easy to deploy in computer and has fast execution, which is useful for application in the real-time operation of EPSs.

5.6 Qualitative characteristics considering the three-phase modeling of the electric network

In the context of three-phase state estimation, the observability and redundancy measurements have some peculiarities. This is not only due to the greater number of state variables to be estimated (complex voltages in the three phases of each EPS bus), but mainly due to the coupling between the phases, which can cause redundancy relations between different phase measurements (Hansen and Debs, 1995).

Another point that must be analyzed, which is related to the qualitative characteristics of the set of measurements considering the three-phase modeling of the network, refers to the treatment of the angular reference. As already mentioned, in the absence of SPMs it is necessary to choose one of the system buses as an angular reference for execution of the single-phase state estimator. Likewise, for the execution of a three-phase state estimator without SPMs, it is necessary to choose an angular reference, which in this case refers to the voltage phase angle in one of the phases of the bus chosen as reference. Consequently, it will be necessary to have measurements to allow the estimation of the voltage phase angles of the other two phases of the reference bus in addition to the voltage magnitudes in all three phases of that bus. However, due to observability problems, especially found in DSs, it is a common practice to assume that the voltage phase angles of the reference bus are balanced, that is, they are offset by 120° between each other. Based on this hypothesis, the set of state variables to be estimated for the reference bus is restricted to the voltage magnitudes in each of its phases.

Although it is quite usual, studies show that using the hypothesis that the phase angles of the voltage phases in the reference bus are balanced can have a great impact on the accuracy of the results depending on the level of unbalance of the electric network under analysis. It should be noted that these studies refer to DSs which, as mentioned in the previous chapter, are intrinsically unbalanced systems.

The methods presented in this chapter for observability analysis and measurement redundancy can be applied in the context of three-phase state estimation. For this, it is enough to apply them to the Jacobian and Gain matrices of the three-phase version of the WLS state estimator, such as that proposed in Zhong and Abur (2002). In this sense, it is important to highlight that in Toyoshima et al. (2012), the method proposed in London et al. (2007) was extended to allow the analysis of observability and redundancy of measurements from the triangular factorization of the Jacobian matrix of the three-phase WLS state estimator proposed in Zhong and Abur (2002).

5.7 Exercises

1. Considering the EPS shown in Fig. 5.13, determine:
   - **(a)** Is it possible to estimate all the state variables? Justify your answer.
   - **(b)** If it is not possible, obtain the observable islands.
   - **(c)** Among the pseudo measurements in Table 5.1, determine the ones necessary to restore the observability of the system (the pseudo measurements must be analyzed considering the order in Table 5.1).
2. Determine, by a topological analysis, the critical measurements and the critical sets of measurements for the EPS shown in Fig. 5.14.

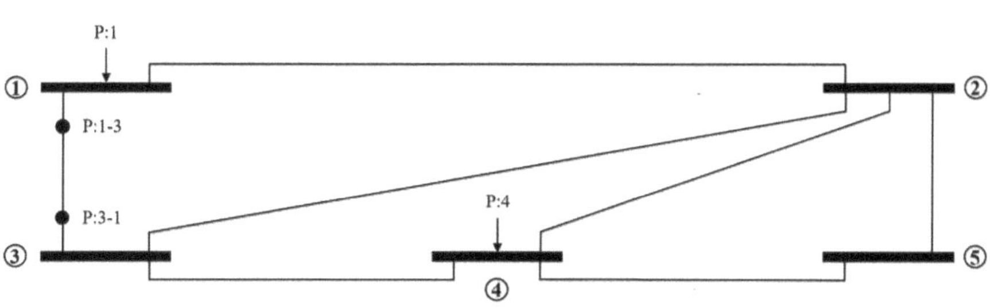

Fig. 5.13 Five-bus test system associated with an active measurement set (•, power flow measurements; →, power injection measurements).

Table 5.1 Available pseudo measurements.

Order	Pseudo measurement
First	P:1–2
Second	P:3–4
Third	P:3

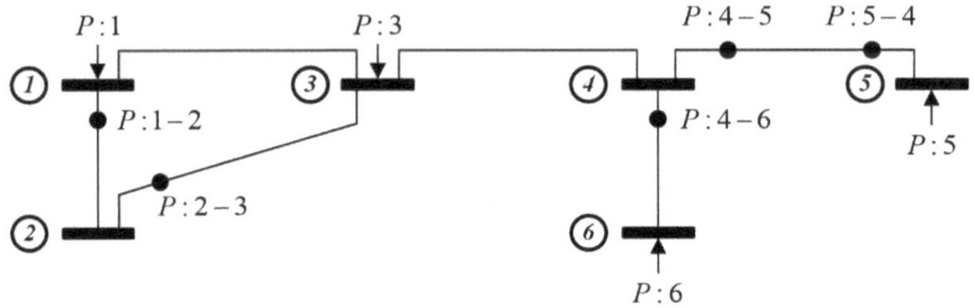

Fig. 5.14 Six-bus test system associated with an active measurement set (•, power flow measurements; →, power injection measurements).

References

Abur, A., Expósito, A., 2004. Power Systems State Estimation: Theory and Implementation. Marcel Dekker, Inc., New York, USA.

Al-Atwan, A., Koglin, H.J., 1998. Two aspects of local redundancy in state estimation. IEE Proc. Gen. Trans. Distrib. 145 (4), 458–462.

Almeida, M.C., Asada, E.N., Garcia, A.V., 2008. Power system observability analysis based on gram matrix and minimum norm solution. IEEE Trans. Power Syst. 23 (4), 1611–1618.

Almeida, M.C., Asada, E.N., Garcia, A.V., 2009. A new method for redundancy analysis of measurements applied to three-phase state estimation. Electr. Power Syst. Res. 79 (1), 234–238.

Ayres, M., Haley, P., 1986. Bad data groups in power system state estimation. IEEE Trans. Power Syst. 1 (3), 1–7.

Benedito, R.A.S., London Jr., J.B.A., Bretas, N.G., 2009. A Unified Algorithm for Observability and Redundancy Analysis. IEEE Powertech Conference, Bucharest, Romania, pp. 1–7.

Bretas, N., 1996. Network observability: theory and algorithms based on triangular factorization and path graph concepts. IEE Proc. Gen. Trans. Distrib. 143 (1), 123–128.

Bretas, N.G., London Jr., J.B.A., Alberto, L.F.C., Bretas, A.S., 2005. A topological approach to the identification of critical measurements in power-system state estimation. IEEE Trans. Circ. Syst. I Regul. Pap. 52 (1), 139–147.

Brueni, D., Heath, L., 2005. The PMU placement problem. SIAM J. Discret. Math. 19 (3), 744–761.

Castillo, E., Conejo, A.J., Pruneda, R.E., Solares, C., 2006. Observability analysis in state estimation: a unified numerical approach. IEEE Trans. Power Syst. 21 (2), 877–886.

Celik, M., Liu, W., 1995. An incremental measurement placement algorithm for state estimation. IEEE Trans. Power Syst. 10 (3), 1698–1703.

Chen, J., Abur, A., 2006. Placement of PMUs to enable bad data detection in state estimation. IEEE Trans. Power Syst. 21 (4), 1608–1615.
Clements, K., Wollember, B., 1975. An Algorithm for Observability Determination in Power System State Estimation. IEEE PES Summer Meeting, San Francisco, USA, pp. 447–453.
Clements, K., Krumpholz, G., Davis, P., 1981. Power system state estimation residual analysis: an algorithm using network topology. IEEE Trans. Power App. Syst. 100 (4), 1779–1787.
Contaxis, G., Korres, G., 1988. A reduced model for power system observability: analysis and restoration. IEEE Trans. Power Syst. 3 (4), 1411–1417.
Coser, J., Costa, A.S., Rolim, J.G., 2006. Metering scheme optimization with emphasis on ensuring bad-data processing capability. IEEE Trans. Power Syst. 21 (4), 1903–1911.
Costa, A.S., Piazza, T.S., Mandel, A., 1990. Qualitative methods to solve qualitative problems in power system state estimation. IEEE Trans. Power Syst. 5 (3), 941–949.
Coutto Filho, M.B., Souza, J.C.S., Schilling, M.T., 2007. Handling critical data and observability. Electr. Power Comp. Syst. 35 (5), 553–573.
Crainic, E.D., Horisberger, H.P., Do, X.D., Mukhedkar, D., 1990. Power network observability: the assessment of the measurement system strength. IEEE Trans. Power Syst. 5 (4), 1267–1285.
Falcao, D., Arias, M., 1994. State estimation and observability analysis based on echelon forms of the linearized measurement models. IEEE Trans. Power Syst. 9 (2), 979–987.
Garcia, A., Monticelli, A., Abreu, P., 1979. Fast decoupled state estimation and bad data processing. IEEE Trans. Power Syst. 98 (5), 1645–1652. https://doi.org/10.1109/TPAS.1979.319482.
Hansen, C., Debs, A., 1995. Power system state estimation using three-phase models. IEEE Trans. Power Syst. 10 (2), 818–824.
Korres, G.N., Contaxis, G.C., 1991a. A reduced model for bad data processing in state estimation. IEEE Trans. Power Syst. 6 (2), 550–557.
Korres, G.N., Contaxis, G.C., 1991b. Identification and updating of minimally dependent sets of measurements in state estimation. IEEE Trans. Power Syst. 6 (3), 999–1005.
Krumpholz, G., Clements, K., Davis, P., 1980. Power system observability: a practical algorithm using network topology. IEEE Trans. Power App. Syst. 99 (4), 1534–1542.
London Jr., J.B.A., Brito, G., Bretas, N.G., 2003. In: Method for meter and RTU placement for state estimation purposes. IEEE Powertech Conference, Bologna, Italy. vol. 1, pp. 1–6.
London Jr., J.B.A., Bretas, A.S., Bretas, N.G., 2004. Algorithms to solve qualitative problems in power system state estimation. Int. J. Electr. Power Energy Syst. 26 (8), 583–592.
London Jr., J.B.A., Alberto, L.F.C., Bretas, N.G., 2007. Analysis of measurement-set qualitative characteristics for state-estimation purposes. IET Gener. Transm. Distrib. 1 (1), 39–45.
London Jr., J.B.A., Piereti, S.A.R., Benedito, R.A.S., Bretas, N.G., 2009. Redundancy and observability analysis of conventional and PMU measurements. IEEE Trans. Power Syst. 24 (3), 1629–1630.
Mili, L., van Cutsem, T., Ribbens-Pavella, M., 1984. Hypothesis testing identification: a new method for bad data analysis in power system state estimation. IEEE Power Eng. Rev. 4 (11), 31–32.
Milosevic, B., Begovic, M., 2003. In: Nondominated sorting genetic algorithm for optimal phasor measurement placement. IEEE PES General Meeting, Toronto, Canada. vol. 2, pp. 639–645.
Monticelli, A., 1999. State Estimation in Electric Power Systems: A Generalized Approach. Kluwer Academic Publishers, Boston, USA.
Monticelli, A., Wu, F., 1985. Network observability: identification of observable islands and measurement placement. IEEE Power Eng. Rev. 5 (5), 32.

Monticelli, A., Wu, F., 1986. Observability analysis for orthogonal transformation based state estimation. IEEE Trans. Power Syst. 1 (1), 201–206.
Mori, H., Tsuzuki, S., 1991. A fast method for topological observability analysis using a minimum spanning tree technique. IEEE Trans. Power Syst. 6 (2), 491–500.
Müller, H.H., Castro, C.A., 2016. Genetic algorithm-based phasor measurement unit placement method considering observability and security criteria. IET Gen. Trans. Distrib. 10 (1), 270–280.
Pussell, L., Trimble, S.Y., 1991. Gram-Schmidt orthogonalization by Gauss elimination. Am. Math. Mon. 98 (6), 544–549.
Quintana, V.H., Costa, A.S., Mandel, A., 1982. Power system topological observability using a direct graph-theoretic approach. IEEE Trans. Power App. Syst. 101 (3), 617–626.
Souza, J.C.S., Coutto Filho, M.B., Schilling, M.T., Capdeville, C., 2005. Optimal metering systems for monitoring power networks under multiple topological scenarios. IEEE Trans. Power Syst. 20 (4), 1700–1708.
Sttot, B., Alsac, O., 1974. Fast decoupled load flow. IEEE Trans. Power Syst. 93, 859–869. https://doi.org/10.1109/TPAS.1974.293985.
Tinney, W.F., Brandwajn, V., Chan, S.M., 1985. Sparse vector methods. IEEE Trans. Power App. Syst. 104 (2), 295–301.
Toyoshima, D., Castillo, M.R.C., Fantin, C.A., London Jr., J.B.A., 2012. In: Observability Analysis and Identification of Critical Measurements on Three-Phase State Estimation. IEEE PES T&D, Orlando, USA, pp. 1–7.
Van Cutsem, T., 1985. Power system observability and related functions derivation of appropriate strategies and algorithms. Int. J. Electr. Power Energy Syst. 7 (3), 175–187. https://doi.org/10.1016/0142-0615(85)90047-X.
Vigliassi, M.P., London Jr., J.B.A., Bretas, N.G., Delbem, A.C.B., 2009. In: Metering System Planning for State Estimation Via Evolutionary Algorithm and HΔ Matrix. IEEE Powertech Conference, Bucharest, Romania, pp. 1–6.
Vigliassi, M.P., Massignan, J.A.D., Delbem, A.C.B., London Jr., J.B.A., 2019. Multi-objective evolutionary algorithm in tables for placement of SCADA and PMU considering the concept of Pareto frontier. Int. J. Electr. Power Energy Syst. 106, 373–382.
Wu, F., Monticelli, A., 1985. Network observability: theory. IEEE Trans. Power App. Syst. 104 (5), 1042–1048.
Zhong, A., Abur, A., 2002. In: Effects of nontransposed lines and unbalanced loads on state estimation. IEEE PES Winter Meeting, New York, USA. vol. 2, pp. 975–979.
Zhu, J., Abur, A., 2007. In: Effect of Phasor Measurements on the Choice of Reference Bus for State Estimation. IEEE PES General Meeting, Tampa, USA, pp. 1–5.

CHAPTER 6

Gross error processing in measurements

6.1 Introduction

The processing of GEs in analog measurements consists of the processes of detection, identification, and elimination of measurements having GE, or the effect of these measurements. Therefore if any measurement is identified as a carrier of GE in Step 4, a further execution of Step 3, i.e., the state estimation itself, will be required. These two steps will then be performed iteratively until no GE is detected in Step 4, ending the entire PSSE process. The state variables estimated in Step 3 will then be used to define the current system operating condition.

It is important to remember, as presented in Chapter 2, that GEs are not grotesque errors, that is, obvious errors that can be detected and eliminated before the PSSE process by simple plausibility checks of the measured values (task performed by prefiltration).

This chapter discusses the processing of GEs through the analysis of WLS static state estimator residuals, since it is the most used state estimator in practice and most studied in academia. In this context, it is worth remembering that the processing of GEs from the analysis of WLS estimator residuals may fail in some situations, mentioned in Chapter 1, but should be remembered as follows:

- GEs associated with low redundancy measurements (CMs or belonging to CSMs);
- Interactive multiple GEs;
- GEs in measurements that have the characteristic of highly influential, that is, attract the convergence of the state estimation process, traditionally called leverage point measurements (Monticelli, 2000).

These limitations led to the development and application of statistically more robust state estimators in EPS, such as the WLMS estimator presented in Chapter 4. However, due to the practical unfeasibility of the WLMS estimator and given the simplicity of the formulation and the ease of computer implementation of the WLS estimator, several research works have been

Cyber-Physical Power Systems State Estimation
https://doi.org/10.1016/B978-0-323-90033-1.00009-3

conducted to improve GE processing through WLS state estimator residual analysis.

In Monticelli and Garcia (1983), the so-called $\hat{b}$ test is used together with the largest normalized residual test to improve the identification process of possible GEs by analyzing the WLS estimator residuals.

Another way to identify GEs through the WLS estimator was proposed in Mili et al. (1984), known as Hypothesis Testing Identification (HTI test), which is based on a statistical analysis concerning the estimated error values rather than analysis of normalized residuals. The efficiency of the HTI test will depend on the small set of measurements chosen, also called the suspected measurement set (set of measurements suspected to have GE). Consequently, this set should be chosen to include all possible error-causing measurements, using the largest standardized residual as a criterion. It should be noted, however, that the choice of this suspected set through the largest standardized residual may be compromised as there is a possibility that one or more GE-carrying measurements may have low normalized residual values (Abur and Expósito, 2004) and, therefore bad estimates will be obtained for the error values.

From the geometric analysis of the WLS estimator residuals (London et al., 2004), it was shown that, in certain redundant measurements, the difficulty in detecting GEs is a structural problem of the detection process from the analysis of WLS estimator residuals. Following this path (Benedito et al., 2014), it was developed an index that allows the classification of measurements according to their characteristics of not reflecting much of their errors in the WLS estimator residuals. This index was called the Undetectability Index (UI). Measurements with the highest UI are those whose GEs are most difficult to detect by analyzing WLS state estimator residuals. This is because the errors of these measurements are somewhat masked, i.e., they are not apparent in the WLS estimator residuals. Because of this, the processing of GEs, either by using the largest normalized residual test or the so-called $\hat{b}$ test (Monticelli and Garcia, 1983), may fail to detect GEs in measurements with high UI indices.

As shown in Benedito et al. (2014), through the UI index it is possible to obtain an estimate of the magnitude of the measurement errors. Thus instead of eliminating the measurement with error, it becomes possible to correct its value and then reestimate the state variables, as proposed in Monticelli and Garcia (1983), based on the so-called $\hat{b}$ test.

Using also a geometric analysis of the WLS state estimator residuals (Bretas et al., 2009, 2011, 2013), new methods for processing GEs were developed. The results obtained from the application of these methods are quite interesting, representing an alternative path for GE processing. Thus these methods and all necessary theoretical background will be dealt with in a special way in Chapter 7.

Seeking an alternative path (Coutto Filho et al., 2014), it was proposed to use SPMs, together with SCADA measurements, to improve GE processing through WLS estimator residual analysis.

In Pesteh et al. (2019) an innovative state estimator is provided that deviates from the orthodox approach in two combined perspectives: the association of modeling the cost function based on generalized error correntropy instead of least squares, with an interior point method solver instead of a Gauss–Newton algorithm. This blend achieves a remarkable level of accuracy in detecting, identifying, and quantifying GEs. The performance of proposed state estimator with the data set contaminated with up to five simultaneous GEs is assessed with confusion matrices, identifying false and missed detections. The superiority of the new method over the classical largest normalized residual test is confirmed at a 99% confidence level in a battery of tests. Its ability to address cases where GEs fall on CMs, CSMs, or leverage points is also confirmed at the same level of confidence.

It is important to highlight that the interest for more robust techniques for detection and identification of GEs has greatly increased due to the possibility that the information provided to state estimators could suffer cyberattacks (Ashok et al., 2018; Chaojun et al., 2015). Initially, it was believed that the techniques for processing GEs based on the analysis of WLS estimator residuals would be able to detect the presence of measurements corrupted by these attacks. However, Liu et al. (2011) demonstrated that a specific type of attack on information processed by the WLS estimator, called false data injection attacks, may go unnoticed by the analysis of WLS estimator residuals. In this sense, it is important to highlight that some studies indicate that the analysis of the residuals fails as a result of these attacks causing a type of GE that is not reflected in the measurement residuals (Chakhchoukh and Ishii, 2015), as well as the GEs with high measurement UI.

The following sections of this chapter will present definitions and the necessary theoretical basis for the understanding of the classical methodologies developed for GE processing from the WLS estimator residuals analysis. After presenting these methodologies, this chapter also provides some examples and exercises to enable a better understanding of the information presented.

6.2 Classification of gross errors

Errors in analog measurements provided to the state estimator can be classified in many ways, depending on the type, location, and number of measurements that have errors. They can be broadly classified as (Abur and Expósito, 2004):

- Single gross error (or singe bad data): only one of the measurements in the entire system has GE;
- Multiple gross errors: more than one measurement has GEs.

Multiple errors can appear in measurements whose residuals are strongly or poorly correlated. Measurements with strongly correlated residuals are those whose errors affect the estimated value of others significantly and may cause one correct measurement to have a high residual due to the effect of the other when it has GE. Estimated values of measurements with poorly correlated residuals are not significantly affected by each other's errors. When the residuals of the measurements are strongly correlated (interactive), their errors may or may not be conforming (Abur and Expósito, 2004).

Nonconforming interactive multiple GEs are errors whose effect is not "masked" in the residual from the GE-carrying measurements. Therefore measurements with GEs will present high normalized residuals. In the case of multiple interactive and conforming GEs, the effect of GE is "masked" in the measurement residuals, causing them to behave similarly to measurements without GE. Thus GE-carrying measurements will have small normalized residuals and measurements without GEs can be identified as GE-carrying (Mili et al., 1984).

Fig. 6.1 illustrates how GEs can be traditionally classified.

6.3 Redundancy vs GE processing

Power systems may contain various types of meters with different properties, which affect the response of the PSSE process as a function not only of the measured values but also of the quantity and location where they are installed. That is, the number and location of meters installed in an EPS affect the accuracy of the estimates obtained. Such happens because this information defines the local redundancy level of the measurements that will be processed by the state estimator. Depending on the local redundancy level, the processing of GEs by analysis of the WLS estimator residuals becomes unfeasible.

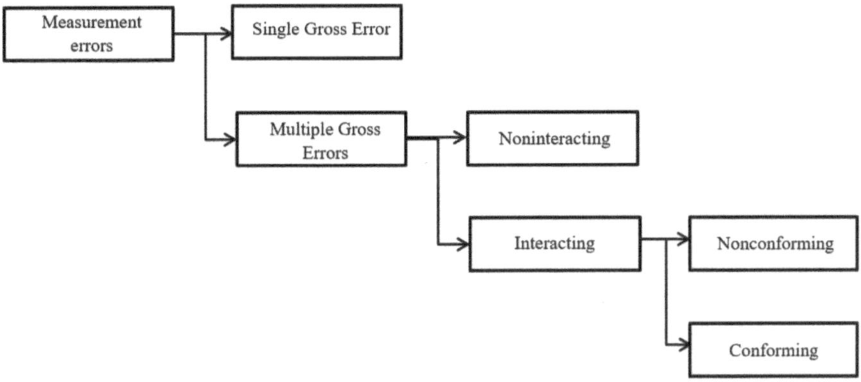

Fig. 6.1 Error classification in analog measurements.

Given the previous, before presenting the methods for processing GEs, it is important to remember some definitions related to local redundancy presented in Chapter 5, but at this time with an emphasis on the processing of GEs rather than the problem of observability. Therefore in this context, measurements may fall into one or more of the following categories (Clements and Davis, 1986):

- Critical Measurement (CM): A measurement is called critical when, once removed from an observable system (or an observable island), it causes the system (or island) to become unobservable. The column of the residual covariance matrix (Ω), which will be presented later in this chapter, corresponding to a CM will be null. Moreover, the residual of a CM will always be zero, which does not allow error detection in this measurement through techniques based on residual analysis.
- Redundant Measurement: A redundant measurement is a measurement that is not critical. Only redundant measurements have nonzero residuals.
- Critical Pair: A critical pair (or p-critical set with $p=2$ as defined in Chapter 5) is composed of two redundant measurements, where the suppression of these measurements makes an observable system (or observable island) unobservable. Suppressing only one measurement of a critical pair makes the other measurement critical, but the system (or island) remains observable.
- Critical Set of Measurements (CSM): It is a set of measurements formed by redundant measurements, in which the elimination of any measurement of it makes the other remaining measurements critical. According

to Mili et al. (1984), the error in a measurement of a CSM can be detected but not identified, since the normalized residuals of measurements from the same CSM have equal magnitude.

Taking into consideration the information presented above, one can reach the following conclusions regarding the WLS estimator associated with the largest normalized residual test (London et al., 2007):

- Any two measurements of a CSM constitute a critical pair of measurements.
- To enable detection of GEs, the local redundancy level of available measurements should be such as to ensure the absence of CMs.
- To enable the identification of GEs, the local redundancy level of available measurements should be enough to ensure the absence of CMs and CSMs.

6.4 Properties of measurement residuals

Consider the linearized measurement model equation that was already presented in Chapter 4:

$$\Delta z = H\Delta x + e, \tag{6.1}$$

where $E(e) = 0$ and $cov(e) = R$, which is a diagonal matrix. It is assumed that measurement errors are uncorrelated. Attention should be paid to the fact that measurement residuals can be correlated even if measurement errors are considered independent (Abur and Expósito, 2004). Then, for the WLS estimator, the linearized estimated state vector will be given by

$$\Delta\hat{x} = \left(H^T R^{-1} H\right)^{-1} H^T R^{-1} \Delta z = G^{-1} H^T R^{-1} \Delta z \tag{6.2}$$

and the estimated value of Δz will be

$$\Delta\hat{z} = H\Delta\hat{x} = K\Delta z, \tag{6.3}$$

where $K = HG^{-1}H^T R^{-1}$ and is usually called a projection matrix or hat matrix (Abur and Expósito, 2004).

A simplified idea of the local redundancy of a measurement can be obtained by checking the corresponding values in the lines of matrix K. A diagonal value that is much higher than the off-diagonal elements of K in the same line implies that the estimated value of that measurement is essentially determined by the corresponding measured value, i.e., the local redundancy is low. In addition, the matrix K has the following properties (Abur and Expósito, 2004):

$$K.K.K...K = K; \tag{6.4}$$

$$K.H = H; \tag{6.5}$$

$$(I - K).H = 0, \tag{6.6}$$

where I is the identity matrix of dimension (m x m) and m is the number of measurements.

Now the residuals of the measurements can be expressed as follows:

$$\begin{gathered} r = \Delta z - \Delta\hat{z} = (I - K)\Delta z \\ r = (I - K)(H\Delta x + e)\ (\text{Substituting Eq.}(6.6)) \\ r = (I - K)e = Se. \end{gathered} \tag{6.7}$$

The matrix S, called residual sensitivity matrix, represents the sensitivity of the measurement residuals in relation to the errors of these measurements. This matrix has the following properties:

- It is not a symmetric matrix except if the measurement error covariances are all equal, that is, $R = \kappa.\ I$, where κ is a scalar;
- $S.\ S.\ S...S = S$;
- $S.\ R.\ S^T = S.\ R$.

The WLS state estimator is based on the hypothesis that in the absence of GEs, the errors (or noises) of the measurements follow a zero-mean Gaussian distribution, i.e.:

$$e_i \sim N(0, R_{ii}), \forall\ measurement\ i \tag{6.8}$$

Using the linear relationship between the measurement residuals and their errors given by Eq. (6.7), the mean and covariance, and thus also the probability distribution, of the measurement residuals can be obtained as follows:

$$E(r) = E(S.e) = S.E(e) = 0 \tag{6.9}$$

$$\begin{gathered} \text{cov}(r) = \Omega = E[rr^T] = S.E[ee^T].S^T \\ \text{cov}(r) = S.R.S^T = S.R \end{gathered} \tag{6.10}$$

Therefore: $r \sim N(0, \Omega)$.

Off-diagonal elements of the residual covariance matrix (Ω) can be used to identify measurements that have strong or weak interactions. If $\Omega_{ij} \geq \xi$, then measurements i and j have a strong interaction. Otherwise, these measurements are considered weakly interactive or noninteractive. The threshold ξ depends on the network and how the measurements are distributed in this network (Abur and Expósito, 2004).

The residual covariance matrix has some interesting properties that will be used in the process of identifying GEs. These properties are presented below:

1. Ω is a real and symmetrical matrix;
2. $\Omega_{ij}^2 \leq \Omega_{ii} \cdot \Omega_{jj}$;
3. $3.\Omega_{ij} \leq \frac{\Omega_{ii} + \Omega_{jj}}{2}$.

6.5 Detection and identification of measurements with gross errors

Detection is the process to determine whether the measurement set contains any measurements with GE. Identification is the procedure of finding which measurements have GEs. Detection and identification of GEs depend on the measurement set configuration in each EPS, i.e., it depends on the local redundancy of the measurements.

GEs can be detected if removing the corresponding measurement does not make the system unobservable. In other words, a GE in a CM cannot be detected. If only one measurement in the set contains GE, this measurement can be identified if and only if it is not a CM nor in a CSM. If both above conditions are met, it becomes possible to detect and identify single GEs by the methods that will be introduced in the next sections. It is noteworthy that, as presented in Bretas et al. (2009); Zhao and Mili (2018); and Massignan et al. (2020), these methods may fail in certain cases of single GEs.

The commonly used methods for detecting GEs for the WLS estimator are the $J(x)$ index test, based on the chi-square distribution (χ^2) analysis, and the largest normalized residual test.

6.5.1 Chi-square distribution and $J(x)$ index test

If x_i ($i = 1, \ldots, m$) are independent random variables that follow a standard normal distribution[a], then the random variable $y = \sum_{i=1}^{m} x_i^2$ will follow a χ_m^2 distribution (chi-square distribution with m degrees of freedom). It can be shown that χ_m^2 has mean m, variance 2 m, and that, as the number

[a] Normal distribution with zero mean and unitary variance, i.e., $x_i \sim N(0, 1)$.

of degrees of freedom (m) increases, this distribution approaches a normal distribution. If the random variables x_i are restricted to a set of N independent equations, then y will follow a χ^2_{m-N} distribution with the number of degrees of freedom reduced by the number of constraints (Monticelli, 1999). Fig. 6.2 illustrates the probability density function of χ^2_m for three different cases: $m = 2$, $m = 4$, and $m = 8$.

Assuming that the errors of measurements e_i ($i = 1, \ldots, m$) are independent and that each has zero mean and variance σ_i^2, i.e., $e_i \sim N(0, \sigma_i^2)$, then the index $J(\hat{x})$ can be given by

$$J(\hat{x}) = \sum_{i=1}^{m} \frac{[z_i - h_i(\hat{x})]^2}{\sigma_i^2}. \tag{6.11}$$

Thus the index $J(\hat{x})$ has a distribution χ^2_{m-N}, that is, a chi-square distribution with $m - N$ degrees of freedom, where m is the number of measurements and N is the number of state variables.

Under these conditions, the expected value of $J(\hat{x})$ and its variance are given by

$$\begin{aligned} E[J(\hat{x})] &= m - N \\ E\{[J(\hat{x}) - (m - N)]^2\} &= 2(m - N) \end{aligned} \tag{6.12}$$

Note that for an observable system with $m = N$, all residuals will be null and $J(\hat{x}) = 0$. When performing state estimation, a $J(\hat{x})$ value is obtained for

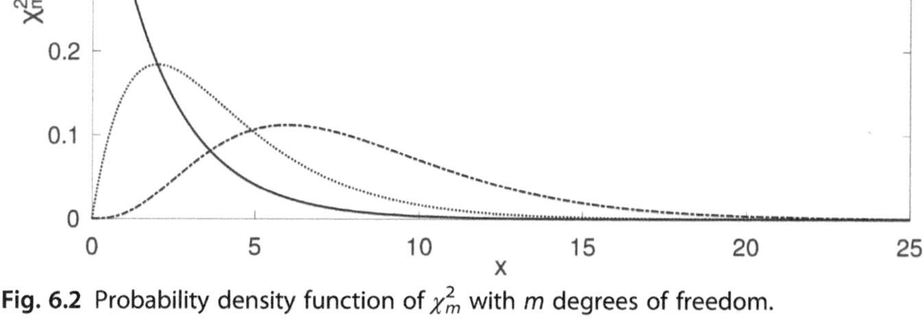

Fig. 6.2 Probability density function of χ^2_m with m degrees of freedom.

the random variable $J(x)$. Based on this estimation, the decision whether this value follows the χ^2_{m-N} distribution must be made. The hypothesis made regarding the behavior of random variables e_i will also be tested (indirectly) for the normal distribution, $N(0, \sigma_i^2)$, so we have a hypothesis test to detect the presence of GEs.

The first step for the GE detection test is to consider a null hypothesis H_0 and an alternative hypothesis H_1, where:

$$\begin{aligned} H_0 &\rightarrow E[J(\hat{x})] = m - n \\ H_1 &\rightarrow E[J(\hat{x})] > m - n. \end{aligned} \tag{6.13}$$

Thus the hypothesis test is done as follows (Monticelli, 1999):

- If $J(\hat{x}) > C$, then reject hypothesis H_0;
- If $J(\hat{x}) \leq C$, then accept hypothesis H_0.

For the hypothesis test, C is a constant to be determined. If α is a test significance level, then the constant C is given by

$$C = \chi^2_{m-N,1-\alpha}. \tag{6.14}$$

This means that if H_0 is true, the probability of $J(\hat{x}) > C$ is α (false alarm probability), where $(1-\alpha)$ can be interpreted as a confidence level of the test, i.e.:

$$\int_0^C f(t)dt = 1 - \alpha. \tag{6.15}$$

The function $f(t)$ represents the probability density function of the distribution χ^2_v, with $v = m - N$ degrees of freedom, and is given by:

$$f(t) = \frac{t^{\frac{v}{2}-1} e^{-\frac{t}{2}}}{2^{\frac{v}{2}} \Gamma\left(\frac{v}{2}\right)}, \tag{6.16}$$

where Γ is the Gamma function. To illustrate the concepts presented with respect to the χ^2 distribution and its properties, in Fig. 6.3, we have the probability density function for the χ^2_8 distribution (chi-square distribution with 8 degrees of freedom): $C = \chi^2_{8;\,0.95} = 15.5$, i.e., $Prob(\chi^2 > 15.5) = 0.05$.

Therefore the objective function $J(x)$ of the WLS state estimator can be used as an approximation to the aforementioned y function. Therefore a GE detection test can be designed based on the properties of the χ^2 distribution.

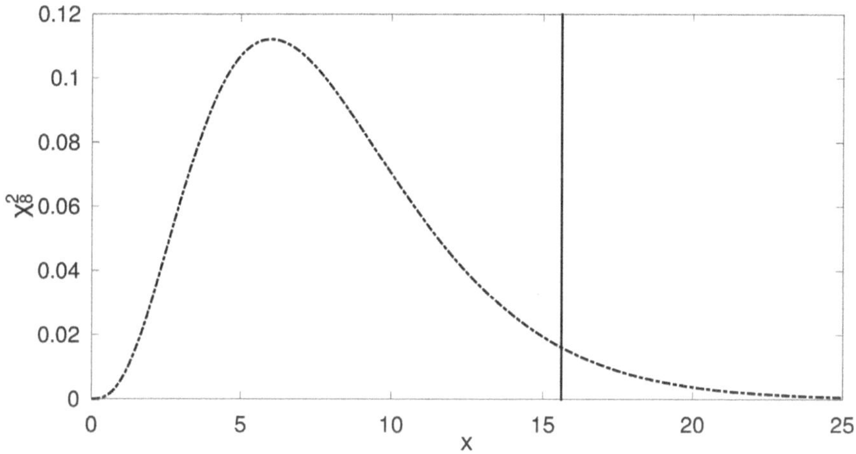

Fig. 6.3 Probability density function for χ_8^2.

The *J*(*x*) index detection test algorithm

- *Step 1*: Solve the WLS estimator and calculate the value of the objective function:

$$J(\hat{x}) = \sum_{i=1}^{m} \frac{[z_i - h_i(\hat{x})]^2}{\sigma_i^2}; \tag{6.17}$$

- *Step 2*: Find in the χ^2 distribution table the value corresponding to a detection with probability $1-\alpha=95\%$ and $(m-N)$ degrees of freedom. Let this value be $\chi^2_{(m-N),\ (1-\alpha)}$, thus

$$1-\alpha = \Pr\left(J(\hat{x}) \leq \chi^2_{(m-N),(1-\alpha)}\right);$$

- *Step 3*: Test if $J(\hat{x}) > \chi^2_{(m-N),(1-\alpha)}$. If so, GE is suspected, otherwise it is assumed that the measurements are free of GE.

6.5.2 Largest normalized residual test

Through analysis of the normalized residuals of the measurements, it is possible to detect and identify the measurements carrying GEs. The normalized value of the residual of a measurement i can be obtained by simply dividing its absolute value by the square root of the corresponding value in the diagonal of the covariance matrix of the estimated residuals:

$$r_i^N = \frac{|r_i|}{\sqrt{\Omega_{ii}}} = \frac{|r_i|}{\sqrt{R_{ii}S_{ii}}}. \tag{6.18}$$

Assuming the hypothesis that the measurement errors (e_i) are independent random variables with zero-mean normal distribution of known variance, it is shown in Monticelli (1999) that the normalized residual vector elements follow a standard normal distribution, that is, $r_i^N \sim N(0,1)$. Thus the largest element in r^N can be compared with a statistical threshold to decide if there is GE. This threshold can be chosen based on the desired level for detection sensitivity.

The importance of residual normalization can be understood by considering that different types of meters generally have different variances, so that the value of the outlier residual for one measurement may be perfectly acceptable for another. The normalization of the residuals puts them in the same reference, thus allowing a fair comparison of their absolute values.

For the case where only one GE-carrying measurement (provided that this measurement is neither a critical nor in a critical set) exists, it can be shown that the largest normalized residual corresponds to the measurement with error given that all other measurements are perfect. This property holds true even in some cases of multiple GEs if the inaccurate measurements are poorly correlated, i.e., as long as they are essentially noninteracting.

Considering the case where the only GE occurs at measurement k, i.e., $e_k \neq 0$, and all the remaining measurements are perfect, $e_j = 0, j \neq k$, and using Eq. (6.7), it can be shown that the normalized residual for the measurement with error k is larger than all error-free measurement residuals as follows:

$$r_j = S_{jk}.e_k, j = 1, \ldots, m.$$

$$r_j^N = \frac{S_{jk}.e_k}{\sqrt{R_{jj}S_{jj}}} = \frac{\Omega_{jk}.e_k}{\sqrt{\Omega_{jj}}.R_{kk}} \leq \frac{\sqrt{\Omega_{jj}}.\sqrt{\Omega_{kk}}.e_k}{\sqrt{\Omega_{jj}}.R_{kk}} = \frac{\sqrt{\Omega_{kk}}.e_k}{R_{kk}} = \frac{S_{kk}.e_k}{\sqrt{R_{kk}}.\sqrt{S_{kk}}} = r_k^N \tag{6.19}$$

Note that the residual covariance matrix property $\Omega_{jk}^2 \leq \Omega_{jj}.\ \Omega_{kk}$ was used. The above inequality becomes an equality if measurements j and k form a critical pair. This is because the columns of the matrix Ω corresponding to these measurements are linearly dependent. Thus when only one of the two measurements of a critical pair carries GE, the normalized residual of that measurement will be the same as one of the other measurements, making it impossible to identify the measurement with GE by the largest normalized residual test. This also occurs when $(p-1)$ measurements from a p-critical set are corrupted with GEs. That is, in that case, it is not possible to identify which measurements carry GEs.

The properties of the normalized residuals presented above can be used to devise a test to detect and identify GE-carrying measurements for the case of single GE in the measurement set. The steps for performing this test, called the Largest Normalized Residual Test (r^N_{max}), are presented below.

The r^N_{max} test algorithm

- *Step 1*: Solve the WLS estimation problem and get the measurement residual vector:

$$r_i = z_i - h_i(\hat{x}); i = 1, \ldots, m.$$

- *Step 2*: Calculate the normalized residual:

$$r_i^N = \frac{|r_i|}{\sqrt{\Omega_{ii}}}; i = 1, \ldots, m.$$

- *Step 3*: Find k so that r_k^N is the largest of all r_i^N, $i = 1, \ldots, m$.
- *Step 4*: If $r_k^N > \beta$, then measurement k is suspected to have GE. Otherwise, stop, there is no suspicion of GE. The β threshold is chosen and usually considered equal to three (Abur and Expósito, 2004).
- *Step 5*: Remove measurement k and return to Step 1.

The largest normalized residual test will perform differently depending on the type of GE and its configuration.

- Single gross error: In the case of single GE, it is possible to identify the inaccurate measurement by testing the largest normalized residual as long as this is not a critical measurement and its removal from the measurement set does not give rise to any critical measurement within the remaining measurements (Abur and Expósito, 2004).
- Multiple Gross Errors: In the case of multiple GEs, these can appear in three ways:

1. Noninteracting: If $S_{ik} \approx 0$, then measurements i and k are said to be noninteracting. In this case, even if the GE appears simultaneously in both measurements, the largest normalized residual test can identify them sequentially, one at a time.
2. Interacting, nonconforming: If S_{ik} is significantly large, then measurements i and k are said to be interacting. However, if errors in measurements i and k are not consistent with each other, testing for the largest normalized residual may still indicate the GE correctly. In other words, they are GEs whose effects are not "masked" in the residuals of the

GE-carrying measurements, so the measurements with GEs will have high normalized residuals, allowing their identification.

3. Interacting, conforming: These are interacting GEs whose effects are "masked" in the residual from the GE-carrying measurements. Thus measurements with GEs will present small normalized residuals, and identification of GEs by testing the largest normalized residual may fail (Abur and Expósito, 2004).

6.5.3 The $\hat{b}$ test

In Monticelli and Garcia (1983), the so-called $\hat{b}$ test is used in association with the largest normalized residual test to improve the GEs identification process. The $\hat{b}$ test is based on an estimate of the GE value in a measurement suspect of having GE. To obtain such an estimate, the residual vector must initially be defined as a function of the so-called residual sensitivity matrix (S) of dimension $m\ x\ m$ (Handschin et al., 1975):

$$r = S.e \tag{6.20}$$

where e is the error vector of analog measurements, as presented in Eq. (6.1). The matrix S, shown in Eq. (6.7), is given by the following equation:

$$S = I - K = I - HG^{-1}H^{T}R^{-1}. \tag{6.21}$$

In order to estimate the GE value in the redundant measurement considered suspect, Eq. (6.20) should be rewritten as follows (Monticelli and Garcia, 1983):

$$r_i = S_{ii}.e_i, \tag{6.22}$$

where measurement i is considered suspect.

From Eq. (6.22) Monticelli and Garcia (1983) shows that the error e_i can be represented as follows:

$$\begin{gathered} e_i = \frac{r_i}{S_{ii}} = \frac{R_{ii}r_i}{\Omega_{ii}} = \sigma_i \frac{\sigma_i r_i}{\Omega_{ii}} \\ e_i = \sigma_i \hat{b}_i, \end{gathered} \tag{6.23}$$

where

$$\hat{b}_i = \frac{\sigma_i r_i}{\Omega_{ii}} = \frac{\sigma_i}{\sqrt{\Omega_{ii}}} \frac{r_i}{\sqrt{\Omega_{ii}}} = \frac{\sigma_i}{\sqrt{\Omega_{ii}}} r_i^N. \tag{6.24}$$

Note 6.1: It should be noted that $\hat{b}_i$ is an estimate of the error of measurement i normalized by its standard deviation.

In view of the above, the existence of GE in the largest standardized residual is verified by the following test:

- If $|\hat{b}_i| > c$, measurement i has GE;
- If $|\hat{b}_i| \leq c$, measurement i has no GE.

In Monticelli and Garcia (1983), the value used and suggested for threshold c was =4. It is noteworthy that because $\hat{b}$ test is based on the analysis of normalized residuals, its performance may be compromised, since there is the possibility of one or more measurements with GEs having low normalized residual values.

The following is the algorithm for processing GEs by the $\hat{b}$ test presented in Monticelli and Garcia (1983):

The $\hat{b}$ test algorithm

Step 1: Determine the state variables $\hat{x}$ using the WLS state estimator;
Step 2: Calculate the normalized residual of each measurement;

$$r_i^N = \frac{|r_i|}{\sqrt{\Omega_{ii}}}; \quad i = 1, \ldots, m. \tag{6.25}$$

Step 3: Find measurement i with the largest normalized residual in modulus;
Step 4: Determine the estimated error $\hat{b}_i$ for measurement i, given by

$$\hat{b}_i = \frac{\sigma_i}{\sqrt{\Omega_{ii}}} \cdot r_i^N \tag{6.26}$$

Step 5: Check if measurement i has GE, i.e.:

- If $|\hat{b}_i| > c$, measurement i has GE, go to step 6;
- If $|\hat{b}_i| \leq c$, measurement i has no GE, end.

Step 6: Remove measurement z_i from the measurement set and return to Step 1.

Note 6.2: Since the removal of the GE-carrying measurement is detrimental to the state estimation process as it results in decreased local redundancy, another option would be to replace the value of this measurement (value with GE) by a "corrected" value as proposed in Garcia et al. (1979). By using $\hat{b}_i$, the "corrected" value of measurement i (z_i^{corr}), identified as carrying GE, can be obtained from the value with GE (z_i^{GE}) by the following equation:

$$z_i^{corr} = z_i^{EG} - \sigma_i.\hat{b}_i. \tag{6.27}$$

6.5.4 Hypothesis test (HTI)

Another way to identify GEs was proposed in Mili et al. (1984), based on a statistical analysis of the estimated error values of the measurements, rather than the analysis of normalized residuals.

According to Mili et al. (1984), the estimated values of measurement errors can be calculated from the following equation:

$$\hat{e} = S^{-1}.r, \tag{6.28}$$

where $\hat{e}$ represents the estimated error vector but does not allow us to estimate the errors of all measurements, since the rank of matrix S is less than the number of measurements m, or more precisely: $rank\ (S) = m - N$. Therefore at most $m - N$ errors in measurements can be calculated using a reduced form of the S matrix.

The efficiency of the method will depend on the small set chosen among the suspected measurements, also called the selected set. Consequently, this set should be chosen to include all possible errors, using as a criterion the largest normalized residuals.

After forming the selected set and calculating the errors of that set ($\hat{e}^S$), the probability distribution for the estimated error should be obtained and then the GEs identified using a hypothesis test similar to the one presented for detection of errors via $J(\hat{x})$ index. Thus more than one measurement with GE can be identified at one time in the occurrence of multiple GEs.

It is important to highlight that the choice of the selected set through the largest standardized residuals compromises the functioning of the HTI test, since, as mentioned before, there is the possibility of one or more GEs measurements presenting low normalized residuals values (Abur and Expósito, 2004).

6.6 Exercises

1. In the continuous current circuit shown in Fig. 6.4, the current and voltage meter readings are given in the following measurement vector:

$$z = [I:1-2\,I:2-1\,I:1-3\,I:3-1\,I:2-3\,I:3-2\,V:1\,V:2\,V:3]^T$$

$$z = [52\,49 - 15\,16 - 82\,80\,11\,6\,14]^T$$

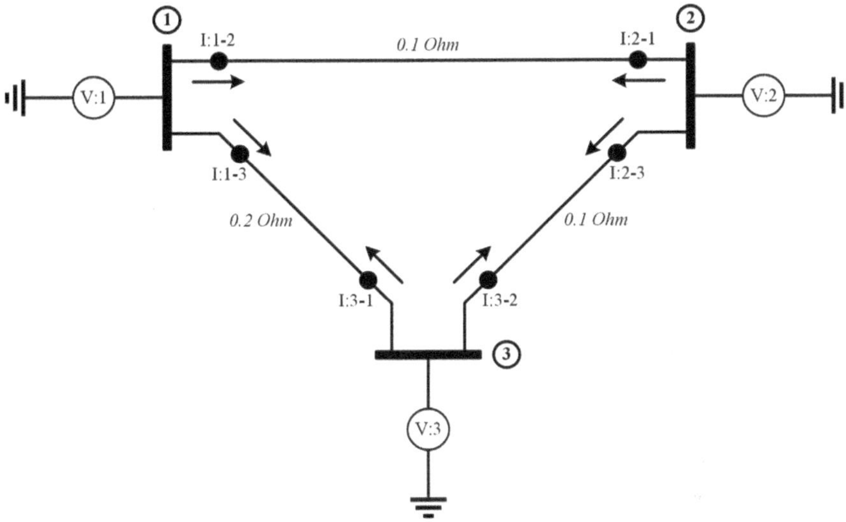

Fig. 6.4 AC three-bus electrical test system.

By analyzing the normalized residuals, check for any GE measurement. If so, identify it for the following situations:

(a) $R^{-1} = diagonal[1\ 1\ 1\ 1\ 1\ 1\ 100\ 100\ 100]$;

(b) $R^{-1} = diagonal[1\ 10\ 1\ 1\ 1\ 1\ 100\ 100\ 100]$.

(c) For item (a), eliminate the measurement in error and reestimate the state variables.

2. For the AC three-bus test system shown in Fig. 6.5 (take bus 1 for angular reference and consider $\varepsilon = 0.001$), whose parameters are shown in Table 6.1, we ask:

(a) Obtain the estimated state variables through the weighted least squares estimator. For this, consider the measurements and their standard deviations presented in Table 6.2:

(b) After obtaining estimates, check for any GE measurement by testing the largest normalized residuals.

3. Consider the IEEE 14 bus system shown in Fig. 6.6. Its bus data are presented in Table 6.3 (note that bus 9 has a shunt element whose value is in pu) and its line data are shown in Table 6.4. With these data, we ask:

(a) Obtain the estimated state variables through the weighted least squares estimator.

(b) After obtaining estimates, check for any GE measurement by testing the largest normalized residual.

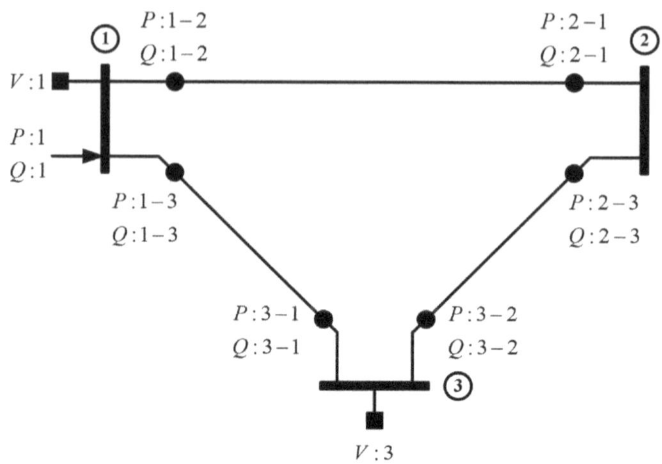

Fig. 6.5 DC three-bus electrical test system.

Table 6.1 Three-bus system parameters illustrated in Fig. 6.5.

Line	g (pu)	b (pu)	b^{sh} (pu)
1–2	4.9991	−15.2630	0.0264
1–3	1.0258	−4.2349	0.0246
2–3	1.7011	−5.1939	0.0173

Table 6.2 Measurement values and their standard deviations.

Measurement	Value (pu)	Standard deviation (σ)
P:1–2	1.5706	0.01047
P:2–1	−1.5276	0.01018
P:1–3	0.7553	0.00503
P:3–1	−0.7277	0.00485
P:2–3	0.4149	0.00276
P:3–2	−0.4039	0.00270
Q:1–2	−0.2033	0.00135
Q:2–1	0.2762	0.00184
Q:1–3	0.0388	0.00026
Q:3–1	0.0219	0.00015
Q:2–3	0.0118	0.00008
Q:3–2	−0.0211	0.00014
P:1	2.3260	0.01551
Q:1	−0.1645	0.00110
V:1	1.0600	0.00707
V:3	1.0200	0.00680

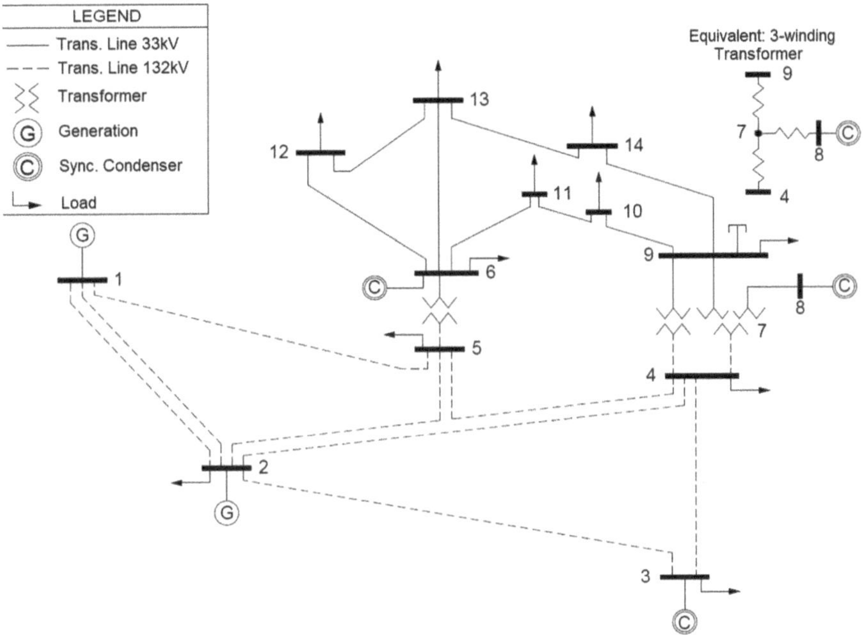

Fig. 6.6 IEEE 14-bus electrical system.

Table 6.3 IEEE 14 bus data (pu).

		Measurements		
Bus (*k*)	b_k^{sh}	P_k	Q_k	**Voltage**
1	0	2.3240	−0.1651	1.0600
2	0	0.1828	0.3084	1.0450
3	0	−0.9420	0.0608	1.0100
4	0	−0.4779	0.0389	1.0180
5	0	−0.0757	−0.0159	1.0200
6	0	−0.1120	0.0519	1.0700
7	0	0	0	1.0620
8	0	0	0.1868	1.0900
9	0.1900	−0.2953	−0.1660	1.0560
10	0	−0.0900	−0.0579	1.0510
11	0	−0.0351	−0.0181	1.0570
12	0	−0.0610	−0.0162	1.0550
13	0	−0.1348	−0.0581	1.0500
14	0	−0.1488	−0.0499	1.0360

Table 6.4 IEEE 14 line data (pu).

						Measurements			
From (k)	To (l)	r_{kl}	x_{kl}	b_{kl}^{sh}	*Tap*	P_{kl}	P_{lk}	Q_{kl}	Q_{lk}
1	2	0.0194	0.0592	0.0264	1	1.5689	−1.5259	−0.2038	0.2765
1	5	0.0540	0.2230	0.0246	1	0.7551	−0.7274	0.0386	0.0221
2	3	0.0470	0.1980	0.0219	1	0.7324	−0.7092	0.0356	0.0160
2	4	0.0581	0.1763	0.0170	1	0.5613	−0.5445	−0.0154	0.0301
2	5	0.0569	0.1739	0.0173	1	0.4151	−0.4061	0.0118	−0.0211
3	4	0.0670	0.1710	0.0064	1	−0.2329	0.2366	0.0448	−0.0484
4	5	0.0133	0.0421	0	1	−0.6118	0.6169	0.1583	−0.1420
4	7	0	0.2091	0	0.9780	0.2809	−0.2809	−0.0968	0.1138
4	9	0	0.5562	0	0.9690	0.1609	−0.1609	−0.0042	0.0173
5	6	0	0.2520	0	0.9320	0.4408	−0.4408	0.1251	−0.0809
6	11	0.0950	0.1989	0	1	0.0736	−0.0731	0.0355	−0.0343
6	12	0.1229	0.2558	0	1	0.0778	−0.0771	0.0251	−0.0236
6	13	0.0661	0.1303	0	1	0.1773	−0.1752	0.0722	−0.0680
7	8	0	0.1762	0	1	0	0	−0.1716	0.1762
7	9	0	0.1100	0	1	0.2808	−0.2808	0.0579	−0.0499
9	10	0.0318	0.0845	0	1	0.0523	−0.0521	0.0423	−0.0419
9	14	0.1271	0.2704	0	1	0.0941	−0.0930	0.0362	−0.0337
10	11	0.0820	0.1921	0	1	−0.0379	0.0380	−0.0159	0.0162
12	13	0.2209	0.1999	0	1	0.0161	−0.0160	0.0075	−0.0074
13	14	0.1709	0.3480	0	1	0.0564	−0.0559	0.0173	−0.0162

Note 1: All system's parameters and measurements are given in p.u.
Note 2: Consider that all the measurements have standard deviations given by: $\sigma_i = \frac{|z_i|}{100}$, being $|z_i|$ the absolute value of measurement i presented in the tables.

References

Abur, A., Expósito, A., 2004. Power Systems State Estimation: Theory and Implementation. Marcel Dekker, Inc., New York, USA.

Ashok, A., Govindarasu, M., Ajjarapu, V., 2018. Online detection of stealthy false data injection attacks in power system state estimation. IEEE Trans. Smart Grid 9 (3), 1636–1646.

Benedito, R.A.S., Alberto, L.F.C., Bretas, N.G., London Jr., J.B.A., 2014. Power system state estimation: undetectable bad data. Int. Trans. Electr. Energy Syst. 24, 91–107.

Bretas, N.G., London Jr., J.B.A., Alberto, L.F.C., Benedito, R.A.S., 2009. Geometrical approaches for gross errors analysis in power systems state estimation. In: IEEE Power-tech Conference, Bucharest, Romania, pp. 1–7.

Bretas, N.G., Bretas, A.S., Piereti, S.A.R., 2011. Innovation concept for measurement gross error detection and identification in power system state estimation. IET Gener. Transm. Distrib. 5 (6), 603–608.

Bretas, N.G., Piereti, S.A.R., Bretas, A.S., Martins, A.C.P., 2013. A geometrical view for multiple gross errors detection, identification, and correction in power system state estimation. IEEE Trans. Power Syst. 28 (3), 2128–2135.

Chakhchoukh, Y., Ishii, H., 2015. Coordinated cyber-attacks on the measurement function in hybrid state estimation. IEEE Trans. Power Syst. 30 (5), 2487–2497.

Chaojun, G., Jirutitijaroen, P., Motani, M., 2015. Detecting false data injection attacks in AC state estimation. IEEE Trans. Smart Grid 6 (5), 2476–2483.

Clements, K., Davis, P., 1986. Multiple bad data detectability and identifiability, a geometric approach. IEEE Power Eng. Rev. PER-6 (7), 73.

Coutto Filho, M.B.D., Souza, J.C.S., Guimaraens, M.A.R., 2014. Enhanced bad data processing by phasor-aided state estimation. IEEE Trans. Power Syst. 29 (5), 2200–2209.

Garcia, A., Monticelli, A., Abreu, P., 1979. Fast decoupled state estimation and bad data processing. IEEE Trans. Power App. Syst. PAS-98 (5), 1645–1652.

Handschin, E., Schweppe, F., Kohlas, J., Flechter, A., 1975. Bad data analysis for power system state estimation. IEEE Trans. Power App. Syst. PAS-94 (2), 329–337.

Liu, Y., Ning, P., Reiter, M.K., 2011. False data injection attacks against state estimation in electric power grids. ACM Trans. Inf. Syst. Secur. 14 (1), 1–33.

London Jr., J.B.A., Bretas, A.S., Bretas, N.G., 2004. Algorithms to solve qualitative problems in power system state estimation. Int. J. Electr. Power Energy Syst. 26 (8), 583–592.

London Jr., J.B.A., Alberto, L.F.C., Bretas, N.G., 2007. Analysis of measurement-set qualitative characteristics for state-estimation purposes. IET Gener. Transm. Distrib. 1 (1), 39–45.

Massignan, J.A.D., London Jr., J.B.A., Vieira, C.S., Miranda, V., 2020. Vulnerability of largest normalized residual test and b-test to gross errors. In: IEEE PES General Meeting, Montreal, Canada, pp. 1–5.

Mili, L., Van Cutsem, T., Ribbens-Pavella, M., 1984. Hypothesis testing identification: a new method for bad data analysis in power system state estimation. IEEE Power Eng. Rev. PER-4 (11), 31–32.

Monticelli, A., 1999. State Estimation in Electric Power Systems: A Generalized Approach. Kluwer Academic Publishers, Boston, USA.

Monticelli, A., 2000. Electric power system state estimation. Proc. IEEE 88 (2), 262–282.
Monticelli, A., Garcia, A., 1983. Reliable bad data processing for real-time state estimation. IEEE Trans. Power App. Syst. PAS-102 (5), 1126–1139.
Pesteh, S., Moayyed, H., Miranda, V., Pereira, J., Freitas, V., Costa, A.S., London Jr., J.B.A., 2019. A new interior point solver with generalized correntropy for multiple gross error suppression in state estimation. Electr. Power Syst. Res. 176, 105937.
Zhao, J., Mili, L., 2018. Vulnerability of the largest normalized residual statistical test to leverage points. IEEE Trans. Power Syst. 33 (4), 4643–4646.

CHAPTER 7

The innovation methodology for error analysis in power systems

7.1 Introduction

In this chapter, the fundamentals of the innovation methodology will be presented, described in detail in Bretas et al. (2009); Bretas and Piereti (2010); Bretas et al. (2013); and Bretas et al. (2017), as well as its application in the processing of errors in the PSSE process. Also presented will be the innovation index (II), which classifies measurements according to their characteristics of "masking" their errors. Error masking of measurements can cause the classic WLS estimator to malfunction, since in general masked errors in measurements are not detected and identified in the estimation process when the classical methods for processing GEs based on the statistical analysis of WLS estimator residuals are used. The II, along with other concepts presented in Bretas et al. (2013), such as composed measurement error (CME) and normalized composed error (CNE), will be used to enable a more efficient analysis of measurement errors.

7.2 Geometric interpretation of measurement errors

As previously stated, when using the WLS state estimator, the objective is to find the best estimate for the vector $\hat{x}$, of dimension N, which minimizes the index $J(x)$. Geometrically, the cost function $J(x)$ is a norm in the $\mathbb{R}^m$ vector space of measurements, induced by the internal product $\langle u, v \rangle = u^T R^{-1} v$:

$$J(x) = \|z - h(x)\|^2_{R^{-1}} = \langle z - h(x), z - h(x) \rangle$$
$$J(x) = [z - h(x)]^T R^{-1} [z - h(x)]. \tag{7.1}$$

Let $\hat{x}$ be the solution to the previously defined minimization problem. Thus the estimated measurement vector is given by $\hat{z} = h(\hat{x})$ and the residual vector is defined as the difference between z and $\hat{z}$, i.e., $r = z - \hat{z}$. The linearization (through Taylor series expansion) of the equation representing the measurement model, given by $z = h(x) + e$, at a given point $\hat{x}$, implies:

Cyber-Physical Power Systems State Estimation
https://doi.org/10.1016/B978-0-323-90033-1.00006-8

$$\Delta z = H\Delta x + e, \tag{7.2}$$

where $H = \frac{\partial h}{\partial x}$ is the Jacobian of h calculated at x^*, $\Delta z = z - h(x^*) = z - z^*$ is the correction in the measurement vector, and $\Delta x = x - x^*$ is the correction in the vector of states.

If the system represented by Eq. (7.2) is observable, then the vector space $\mathbb{R}^m$ of the measurements can be decomposed into a direct sum of two vector subspaces, as follows:

$$\mathbb{R}^m = \mathfrak{R}(H) \oplus [\mathfrak{R}(H)]^{\perp}, \tag{7.3}$$

so that the image of H, given by $\mathfrak{R}(H)$, is an N-dimensional vector subspace in $\mathbb{R}^m$ and $\mathfrak{R}(H)^{\perp}$ is its orthogonal complement, that is, if $u \in \mathfrak{R}(H)$ and $v \in \mathfrak{R}(H)^{\perp}$, then $\langle u, v \rangle = u^T R^{-1} v = 0$.

In the linear state estimation formulation, represented by Eq. (7.2), its solution can be interpreted as the projection of the measurement vector correction Δz into $\mathfrak{R}(H)$. Let K be the linear operator that projects the vector Δz into $\mathfrak{R}(H)$, i.e., $\Delta\hat{z} = K\Delta z$ and let $r = \Delta z - \Delta\hat{z}$ be the residual vector of the measurements for the linearized model. The operator K, which minimizes the norm J (Eq. 7.1), is the one that projects Δz orthogonally on $\mathfrak{R}(H)$, that is, the vector $\Delta\hat{z} = H\Delta\hat{x}$ is orthogonal to the residual vector, in the sense of the inner product $\langle u, v \rangle = u^T R^{-1} v$. More precisely:

$$\Delta\hat{z}, r = (H\Delta\hat{x})^T R^{-1} (\Delta z - H\Delta\hat{x}) = 0. \tag{7.4}$$

Solving this equation for $\Delta\hat{x}$ gives:

$$\Delta\hat{x} = \left(H^T R^{-1} H\right)^{-1} H^T R^{-1} \Delta z. \tag{7.5}$$

Since $\Delta\hat{z} = H\Delta\hat{x}$, the projection matrix K will be the idempotent matrix:

$$K = \mathrm{H}\left(\mathrm{H}^\mathrm{T} \mathrm{R}^{-1} \mathrm{H}\right)^{-1} \mathrm{H}^\mathrm{T} \mathrm{R}^{-1}. \tag{7.6}$$

In PSSE literature, matrix K is usually called Hat matrix or projection matrix, as presented in Chapter 6. Therefore geometrically, the WLS state estimator can be viewed as a projection matrix K applied to the measurement vector mismatch Δz, as shown in Fig. 7.1, where $\Delta\hat{z}$ is its estimated value.

Another way to visualize the relationship between the error of a measurement and the span of the columns of the Jacobian matrix ($\mathfrak{R}$(H)) is obtained by decomposing the vector space of the measurements into a direct sum of $\mathfrak{R}$(H) and $\mathfrak{R}(H)^{\perp}$. Thus it is possible to decompose the measurement

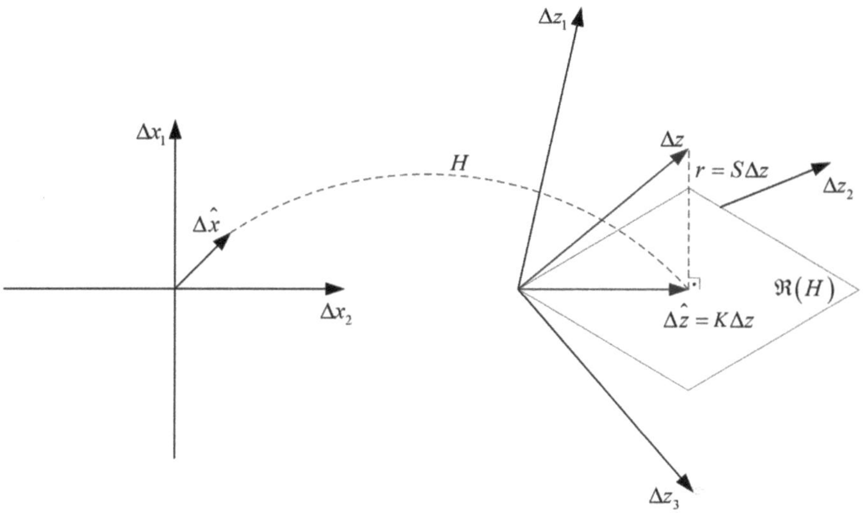

Fig. 7.1 Geometric interpretation of operator K acting on vector Δz.

error vector "e" into two components: detectable and undetectable, as follows:

$$e = Ke + (I - K)e. \tag{7.7}$$

Denoting e_U and e_D, respectively, as the undetectable and detectable components of the measurement error vector "e", given by.

$$\begin{aligned} e_U &= Ke \\ e_D &= (I - K)e. \end{aligned} \tag{7.8}$$

The measurement error vector can be rewritten as.

$$e = e_U + e_D, \tag{7.9}$$

where $e_U \in \Re(H)$ and $e_D \in \Re(H)^{\perp}$, so that.

$$\|e\|_{R^{-1}}^2 = \|e_U\|_{R^{-1}}^2 + \|e_D\|_{R^{-1}}^2. \tag{7.10}$$

- **Proposition:** The residual vector r associated with WLS SE is not affected by the undetectable component (e_U) of the measurement error vector.
- **Demonstration:** From the WLS estimator equation, we have that

$$\Delta\hat{x} = G^{-1}H^T R^{-1}\Delta z. \tag{7.11}$$

The estimated value of Δz is also given by.

$$\Delta\hat{z} = H\Delta\hat{x} = K\Delta z. \tag{7.12}$$

As presented in Section 6.4, the projection matrix has the following property:

$$(I - K)\,H = 0, \tag{7.13}$$

Therefore the residual of the measurements can be expressed as.

$$r = \Delta z - \Delta\hat{z} = (I - K)\Delta z$$
$$r = (I - K)(H\Delta x + e) \ \text{(SubstitutingEq.7.13)}$$
$$r = (I - K)e = e_D. \tag{7.14}$$

q.e.d.

From the analysis of this proposition, it can be concluded that small normalized residuals do not imply small measurement errors, since the measurement residual is not affected by the undetectable component of the errors. Then the measurement residual does not change if an extra error that is in the vector space of $\Re(H)$ is added to the measurement vector. In other words, when the value of the largest normalized residual is not sufficient to detect GE, it can be safely stated that only the detectable error component in all available measurements is small. As a result, testing for the largest normalized residual can often fail to detect GE for measurements that have a large undetectable component.

The above proposition indicates that the difficulty, or even impossibility, of detecting GEs in some measurements is due to the high ratio between undetectable and detectable components of its errors, i.e., $\|e_U\|_{R^{-1}}/\|e_D\|_{R^{-1}}$, where $\|e_U\|_{R^{-1}}$ is the norm of e_U defined by the inner product $\langle u, v\rangle = u^T R^{-1} v$.

Thus the ratio between the error components was defined as an *Innovation Index* (*II*) (Bretas et al., 2011), that is, a contribution that measurement brings to the PSSE process. So, suppose there is a simple error in the ith measurement, that is.

$$e_i = b\delta_i, \tag{7.15}$$

where the vector $\delta_i = \left[0, \ldots, 0, \overbrace{1}^{} i, 0, \ldots, 0\right]^T$ and b is any scalar representing the magnitude of the error. As shown in Eq. (7.8), detectable and undetectable components are calculated by.

$$
\begin{aligned}
e_U &= Ke \\
e_D &= (I-K)e.
\end{aligned} \tag{7.16}
$$

The II is defined for the ith measurement as (Bretas et al., 2011):

$$
II_i = \frac{\|e_D^i\|_{R^{-1}}}{\|e_U^i\|_{R^{-1}}} = \frac{\sqrt{(e_D^i)^T R^{-1}(e_D^i)}}{\sqrt{(e_U^i)^T R^{-1}(e_U^i)}} \tag{7.17}
$$

The above equation can be expressed as a function of matrices K and $(I-K)$, that is.

$$
\begin{aligned}
II_i &= \frac{\sqrt{((I-K)e_i)^T R^{-1}((I-K)e_i)}}{\sqrt{(Ke_i)^T R^{-1}(Ke_i)}} \\
II_i &= \frac{\sqrt{((I-K)b\delta_i)^T R^{-1}((I-K)b\delta_i)}}{\sqrt{(Kb\delta_i)^T R^{-1}(Kb\delta_i)}} \\
II_i &= \frac{b\sqrt{((I-K)\delta_i)^T R^{-1}((I-K)\delta_i)}}{b\sqrt{(K\delta_i)^T R^{-1}(K\delta_i)}}
\end{aligned} \tag{7.18}
$$

From this representation for II and, using mathematical demonstration, it is shown that the numerator and denominator of the above equation respectively result in (see Appendix B):

$$
\begin{aligned}
\left\|e_D^i\right\|_{R^{-1}} &= b\sigma_i^{-1}\sqrt{(1-K_{ii})} \\
\left\|e_U^i\right\|_{R^{-1}} &= b\sigma_i^{-1}\sqrt{K_{ii}}
\end{aligned} \tag{7.19}
$$

Thus one can write index II, for the ith measurement, as follows:

$$
II_i = \frac{\sqrt{K_{ii}}}{\sqrt{(I-K_{ii})}}, \tag{7.20}
$$

where K_{ii} is the ith element of the main diagonal of the projection matrix K.

As can be seen from the above equation, II does not depend on the magnitude of error b, but only on the projection matrix K and the sensitivity matrix of the residual S ($S = I - K$). This is because the undetectable error component is directly related to the projection matrix K, or $\mathfrak{R}(H)$, while the detectable component is related to the sensitivity matrix $(I-K)$, or $\mathfrak{R}(H)^{\perp}$. Thus a small value of *II* indicates that a large portion of its error will be "masked" in the state estimation process. Consequently, the error of the measurement will contribute little to the value of its residual (Bretas and Piereti, 2010).

For a clearer understanding of the concept of *II*, an example will be presented with the presence of a measurement with low *II* magnitude and one with high *II* magnitude. For that purpose, consider Fig. 7.2, which represents a simple two-bus example system.

The parameters of this system and the weights of the flow measurements are such that measurement z_1 has low *II* and measurement z_2 has high *II*. In this case, if measurement z_1 has GE, this error will be difficult to detect because

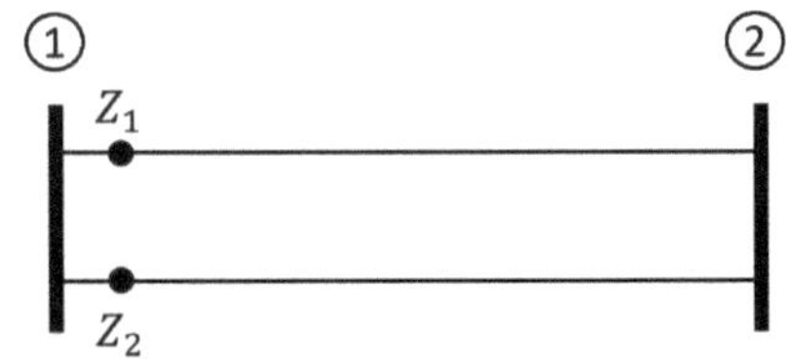

Fig. 7.2 Two-bus system.

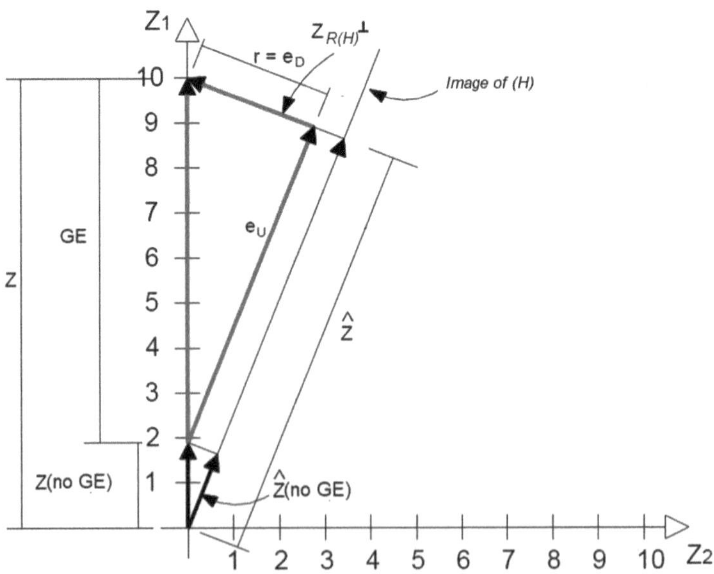

Fig. 7.3 Example of GE in a low *II* measurement (Bretas et al., 2013).

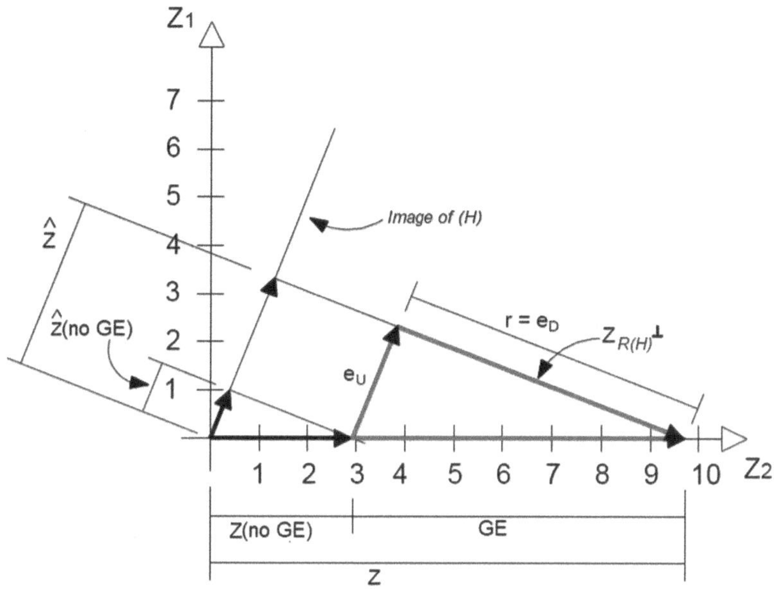

Fig. 7.4 Example of GE in a high *II* measurement (Bretas et al., 2013).

the residual of this measurement will be low (see Fig. 7.3). This is because the detectable component of measurement error z_1 is too small due to its reduced *II*. In contrast, if measurement z_2 has GE, since the error detectable component of this measurement is larger than its undetectable component, its error will be easily detected as it will have a high residual (see Fig. 7.4).

7.3 Innovation concept

The innovation of a measurement (Bretas and Piereti, 2010), in the context of static state estimation, can be defined as the new information that measurement contains in relation to the other measurements of the system, that is, information not contained in other measurements of the measurement set. This definition suggests that innovation is contained in the measurement portion that is independent of the other measurements of the system, that is, the portion that cannot be obtained from linear combinations of the Jacobian lines relative to the previous measurements.

As a result, the new information contained in a measurement is that part of the measurement that is orthogonal to the span of the columns of the Jacobian matrix related to the system's set of measurements. Thus this new information will be the orthogonal projection of the measurement on the Jacobian matrix image, that is, in $\mathfrak{R}(H)^{\perp}$. In case a measurement contains error, its component that is orthogonal to the span of the Jacobian matrix will show the error through the measurement residual; the other component, however, will have its error masked. Thus the masked error vector in the state estimation process is the vector that is contained in the span of the Jacobian matrix (Bretas and Piereti, 2010).

Once the residual of the estimated measurements is known and the error components are orthogonal, it is possible to compose the total vector of measurement errors. That is, for a measurement i, rewriting Eq. (7.10):

$$\|e_i\|_{R^{-1}}^2 = \|e_U^i\|_{R^{-1}}^2 + \|e_D^i\|_{R^{-1}}^2. \tag{7.21}$$

The error vector calculated in this way is called the *composed measurement error* (*CME*). In order to find the masked error and then compose the total error of measurements, the *II* is used:

$$II_i = \frac{\|e_D^i\|_{R^{-1}}}{\|e_U^i\|_{R^{-1}}}. \tag{7.22}$$

Knowing that the vector spaces $\Re(H)$ and $\Re(H)^{\perp}$ are orthogonal to each other, then it is possible to recalculate the compound error of the ith measurement. Thus Eq. (7.21) becomes:

$$\|e_i\|_{R^{-1}}^2 = \frac{1}{II_i^2}\|e_D^i\|_{R^{-1}}^2 + \|e_D^i\|_{R^{-1}}^2$$

$$\|e_i\|_{R^{-1}}^2 = \left(1 + \frac{1}{II_i^2}\right)\|e_D^i\|_{R^{-1}}^2. \tag{7.23}$$

or yet:

$$CME_i^2 = \left(1 + \frac{1}{II_i^2}\right) r_i^2 \Rightarrow CME_i = r_i\sqrt{1 + \frac{1}{II_i^2}}, \tag{7.24}$$

where r_i is the residue of the ith measurement and II_i is the innovation Index of this same measurement, both known quantities. It is important to highlight that $\|e_D\|_{R^{-1}}^2 = r_i$.

If the residual is normalized and then the error composition is made, we obtain the *composed normalized error* (*CNE*), given by:

$$CNE_i = \frac{CME_i}{\sqrt{\Omega_{ii}}} = \frac{r_i}{\sqrt{\Omega_{ii}}}\sqrt{1 + \frac{1}{II_i^2}} = r_i^N\sqrt{1 + \frac{1}{II_i^2}}, \tag{7.25}$$

where r_i^N is the normalized residual of the ith measurement. The *CNE* will be applied later in the correction of the measurement with GE.

7.4 Application of innovation methodology for processing GEs

7.4.1 CME-associated χ^2 distribution

Since measurement errors are random variables, the *CME*-related $J(x)$ index (which is labeled as $J_C(x)$), as in the case of the classical detection test, will also have a chi-square distribution. Therefore the detection of GE using the $J(x)$ index is as presented in Section 6.5, and its value is calculated as follows:

$$J_C(x) = \sum_{i=1}^{m} \left(\frac{CME_i}{\sigma_i} \right)^2, \tag{7.26}$$

7.4.2 New GE detection and identification test

As shown in Bretas and Piereti (2010); Bretas and London (2010); and Bretas et al. (2011), the composed error of the ith measurement can be calculated by.

$$CME_i = r_i \sqrt{1 + \frac{1}{II_i^2}}. \tag{7.27}$$

With this equation, one can obtain the composed error of measurement in its normalized form (CME_i^N), which will be used later in the measurement identification test with GE. The (CME_i^N) is calculated by Bretas et al. (2013):

$$CME_i^N = \frac{r_i}{\sigma_i} \sqrt{1 + \frac{1}{II_i^2}}. \tag{7.28}$$

In order to detect the measurement with GE, the following algorithm is proposed:

Step 1: Perform the state estimation and calculate the residual vector r ($\mathrm{r} = \|e_D\|_{R^{-1}}^2$);

Step 2: Calculate the II of each measurement (Eq. (7.20)) and then calculate the vector of the undetectable error component (e_U), that is, $e_U^i = \left(\frac{1}{II_i}\right) e_D^i$;

Step 3: With these two vectors in hand, calculate the composed error of all measurements (CME_i) using Eq. (7.27);

Step 4: Perform the GE detection test. This step is done in a similar way to the largest normalized residual test, but using the CME^N instead of the measurement normalized residual vector.

In order to identify a measurement with GE, once an error has been detected in the set of measurements, it is necessary to use the following theorem.

Theorem The largest normalized error test (Bretas et al., 2013)

Assuming that all measurements are members of a set of measurements with limited random errors, if one gross error is added to only one of the measurements, that measurement will have the largest error increment of all measurements.

Demonstration: Assume that all measurements are perfect, without any error. The measurement residual vector will be given by:

$$r = (I - K)z = \Omega R^{-1} z, \tag{7.29}$$

which in this situation is zero for all measurements.

Suppose now that an error is added to measurement z_i, with all others remaining at their original values. Consequently, for this new measurement: $z_i^{new} = z_i + b_i \sigma_i$, where index "*new*" represents the new measurement value. Following the same pattern as the demonstration of the classical test of the largest normalized residual presented in Monticelli (1999) (see Chapter 6, Section 6.5.2) gives.

$$r_i^{new} - r_i = b_i \sigma_i^{-1} \Omega_i u_i, \tag{7.30}$$

where $\Omega_i u_i$ is the ith column of the residual covariance matrix (Ω).

Consequently, and following the pattern of the classic demonstration by (Monticelli, 1999), we obtain:

$$\left| r_i^{N,new} - r_i^N \right| \geq \left| r_j^{N,new} - r_j^N \right|, \tag{7.31}$$

or, equivalently.

$$\left| CME_i^{N,new} - CME_i^N \right| \geq \left| CME_j^{N,new} - CME_j^N \right|; j = 1, \ldots, m, \tag{7.32}$$

where index j represents any of the other measurements without error.

It is concluded, then, that the largest increment of error is present in the wrong measurement and this conclusion does not depend on the initial conditions of the measurements. In addition, one can apply generalization that when measurements initially have a random error, a minimum error value will occur in all measurements, so the error will be present in the measurement with the largest CME^N.

q.e.d.

7.4.3 Correction of measurement error

After the error measurement detection and identification process, the measurement correction should be performed applying the following equation (Bretas et al., 2013):

$$z_i^{corr} = z_i^{GE} - CNE_i\sigma_i, \tag{7.33}$$

where for the ith measurement:

z_i^{corr} is the corrected measurement value;
z_i^{GE} is the value of the measurement with error;
CNE_i is its normalized composed error;
σ_i is its standard deviation.

7.4.4 Flowchart of the new innovation methodology for GE processing

The state estimator flowchart associated with the innovation methodology based on the analysis of error components is presented in Fig. 7.5.

Note 7.1: The β threshold chosen for GE detection is the same as the traditional methodology ($\beta = 3$).

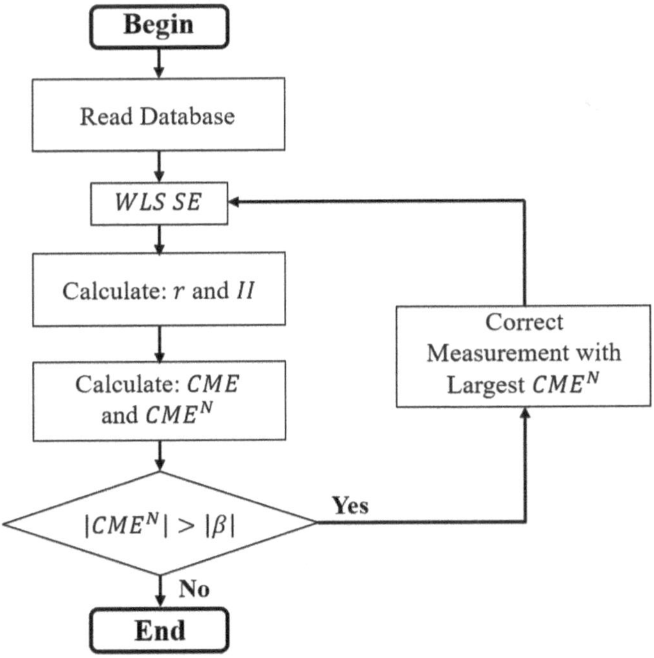

Fig. 7.5 Flowchart of the innovation methodology for processing GEs.

7.5 Application of innovation methodology for parameter and topology error processing

Traditional techniques for processing PEs and TEs are based on the analysis of measurement residual sensitivity. In this work, however, we will use the $\chi 2$ test for error detection, the largest CME^N test for error identification and the CNE as the index for correction of measurement values. If parameters were identified as suspicious of being with error, they will be added as a variable to be estimated.

Methods that increase the vector of state variables by including suspicious parameters as independent variables require changing the computational code of the conventional PSSE process by WLS. Thus the parameters and complex voltages of EPS buses are estimated simultaneously, which is called simultaneous estimation of state and parameters. It is important to note that preliminary execution of the conventional state estimation may be necessary to identify which parameters should be included in the variable vector to be estimated, since, to maintain a reasonable local redundancy level, the number of suspicious parameters included as extra variables should be as small as possible.

One of the first methods concerning the simultaneous estimation of state and parameters, based on the increased state vector via normal equations, was proposed by Allan and Laughton (1973). This method initially had the objective of estimating all elements of the admittance matrix, allowing the estimation of state variables and parameters through the difference between the parameter values provided by the database and their estimated values, and to identify if the system branches are out of operation or not.

In Alsaç et al. (1998), a generalized state estimator was proposed. Its main contribution was the improvement in the process of obtaining information, besides the inclusion of topology errors and parameter errors in the estimation process. The proposed estimator allows modeling of the suspect substations at the bus section level. For this, the conventional state variable vector, consisting of voltage magnitudes and angles, incorporates new state variables, which are the active and reactive power flows through the circuit breakers and branches whose status and parameters will be estimated. New pseudo measurements are also introduced in the model to allow a better distinction between GEs, PEs, and TEs.

In general, it can be argued that the limitation of methods using normal equations is related to observability problems, since rarely the number of measurements available is enough to estimate all state variables. That is particularly relevant for methods that rely on increasing the number of state variables while the set of measurements remains unchanged (Zarco and Expósito, 2000).

Another possibility for estimating the increased vector of state variables is through a recursive algorithm, based on the Kalman filter, initially investigated by Debs (1974). In this approach, the parameters are modeled as constants and the complex voltages updated according to a Markov process. The use of dynamic models allowed the use of recursive estimation, in which a priori information about state variables and parameters are combined with current information (measurements) to update parameter estimation. The limitations of such a formulation are:

- Convergence problems when applied to large systems and/or with many incorrect parameters;
- Considering parameters as constants limits the flexibility of the algorithm since some parameters are variable over time;
- The need to update the PE covariance matrix greatly increases the computational cost of Kalman filter-based methods, especially when the number of suspicious parameters is large (Abur and Expósito, 2004).

Due to its mathematical and implementation complexity, the parameter estimation method based on the Kalman filter will not be presented in detail in this document. If the reader is interested, more information is available in the works of Debs (1974); Slutsker et al. (1996); and Abur and Expósito (2004).

Using orthogonal projections, it was shown in a previous section that the solution of the WLS state estimation problem can be obtained by projecting the measurement vector over space $\mathfrak{R}(H)$. However, as shown in van Cutsem and Quintana (1988) and Zarco and Expósito (2000), when there is any network element with parameter error, the Jacobian matrix (H) is affected. From the above, it can be concluded that, in the presence of PEs, the solution of the WLS sate estimation problem is obtained by projecting the vector Δz over the space $\mathfrak{R}(H_p)$, which corresponds to the electrical system represented with erroneous parameters, and not over the space $\mathfrak{R}(H)$, which corresponds to the system actually in operation.

Fig. 7.6 shows that when there are no types of errors, the vector Δz is close to the space generated by the columns of the Jacobian matrix, since the measurements contain only small random errors and, consequently, the residual vector is small. However, as a consequence of the PEs (or

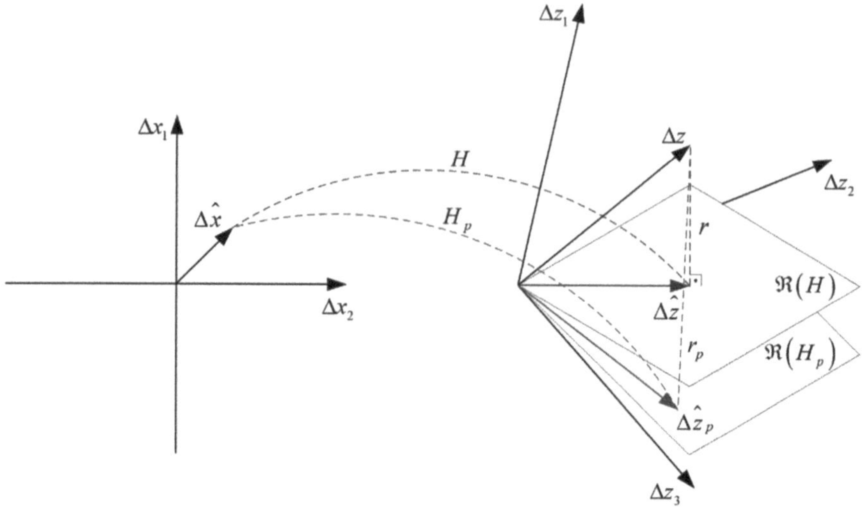

Fig. 7.6 Vector Δz projected over the image space without PEs $\Re(H)$ and with PEs $\Re(H_p)$.

topology, as shown in Simões Costa and Leão (1993)), the space $\Re(H_p)$, relative to the configured system suffers a perturbation, which increases the distance between Δz and the estimation results in a vector. The resulting larger norm of residuals is shown in Fig. 7.6, where the r and r_p residual vectors obtained for the system without PEs and with PEs, respectively, are compared.

Then, the mathematical formulation for estimation of suspicious parameters will be presented by applying the augmented version of the conventional normal equation. The augmented measurement model is given by:

$$z = h(x, p) + e, \tag{7.34}$$

where p is a vector of suspected parameters to be estimated.

The augmented objective function is defined as.

$$J_A(x, p) = \frac{1}{2}[z - h(x, p)]^T R_z^{-1}[z - h(x, p)] + \frac{1}{2}(p_0 - p)^T R_p^{-1}(p_0 - p) \tag{7.35}$$

where p_0 is an a priori value vector (values from the static database) of the suspect parameters.

In order to adapt the WLS estimator computational code, the augmented state variable vector (x_A) and the augmented measurement vector (z_A) are given by.

$$x_A = \begin{bmatrix} x \\ p \end{bmatrix}. \tag{7.36}$$

$$z_A = \begin{bmatrix} z \\ p_0 \end{bmatrix}. \tag{7.37}$$

As previously mentioned, regardless of the technique adopted, these methods increase the number of elements of the Jacobian matrix, since as many extra Jacobian matrix columns are introduced as new variables (suspicious parameters) are added to the state variables vector. The new columns correspond to the partial derivatives of the measurement equations with respect to the new state variables.

The a priori information associated with the new state variables (suspicious parameters) is also added to the measurement vector. Thus as many additional lines will be included in the augmented Jacobian matrix (H_A) as the a priori information added to the measurement vector. The following is the augmented Jacobian matrix H_A resulting from the introduction of a single suspect parameter as a state variable:

$$H_A = \begin{bmatrix} & & & 0 \\ & & & \vdots \\ & & & \oplus \\ & H_x & & \vdots \\ & & & 0 \\ & & & \vdots \\ & & & \oplus \\ 0 \ldots 0 & \ldots & 0 & 1 \end{bmatrix}, \tag{7.38}$$

where H_x corresponds to the Jacobian submatrix that relates the measurements to the conventional state variables (V and θ) and $\oplus$ represents the non-zero elements related with the flow measurement and power injection equations that are dependent on the parameter to be estimated.

In general, the augmented Jacobian matrix has the following structure:

$$H_A = \begin{bmatrix} H_x & H_p \\ 0 & I \end{bmatrix}, \tag{7.39}$$

where I is an identity matrix of dimension equal to the number of parameters to be estimated.

Then the increased gain matrix will be given by Monticelli (1999):

$$G_A = \begin{bmatrix} H_x^T R_z^{-1} H_x & H_x^T R_z^{-1} H_p \\ H_p^T R_z^{-1} H_x & H_p^T R_z^{-1} H_p + R_p^{-1} \end{bmatrix}, \tag{7.40}$$

where R_p is the covariance matrix of the parameters, its inverse is used as a weighting matrix, analogous to the weighting matrix of the measurements R_z^{-1}.

If there is enough redundancy to estimate the suspicious parameters, the value assigned to the parameter standard deviation (σ_p) is very important as it can significantly influence the estimated value $\hat{p}$. Assigning σ_p too large with respect to the standard deviation of measurements σ_i is equivalent to rejecting the prior information p_0 and, therefore, the value of the estimated parameter $\hat{p}$ would be determined solely by the measurements available in the system. Otherwise, if it is assigned a very small value, the influence of the initial parameter value is dominant and there is no way to significantly improve the estimated value. In other words, an estimated value $\hat{p}$ close to p_0 will be obtained regardless of the measurement values (Abur and Expósito, 2004). Through practical tests, Abur and Expósito (2004) suggest that σ_p be given as a function of the mean standard deviation value of the measurements (σ_{mean}), according to the following relation:

$$\frac{\sigma_p}{\sigma_{mean}} \cong 10^3, \tag{7.41}$$

Finally, having the respective augmented versions of the vectors and matrices involved in the state estimation process, the new normal equation can be defined as.

$$G_A \Delta x_A = H_A^T R_A^{-1} \Delta z_A, \tag{7.42}$$

or, in more detail:

$$\begin{bmatrix} H_x^T R_z^{-1} H_x & H_x^T R_z^{-1} H_p \\ H_p^T R_z^{-1} H_x & H_p^T R_z^{-1} H_p + R_p^{-1} \end{bmatrix} \cdot \begin{bmatrix} \Delta x \\ \Delta p \end{bmatrix} = \begin{bmatrix} H_x^T & 0 \\ H_p^T & I \end{bmatrix} \cdot \begin{bmatrix} R_z^{-1} & 0 \\ 0 & R_p^{-1} \end{bmatrix} \cdot \begin{bmatrix} \Delta z \\ p_0 - p \end{bmatrix} \tag{7.43}$$

The error of the ith measurement, e_i, may be of high magnitude due to several factors, such as

- a GE in the ith measurement;
- an error in a parameter associated with the equation of this measurement;

- a topology error on a transmission line containing this measurement;
- a possible combination of the above factors.

The main question is to identify the cause of this high e_i, and what the proposed innovation methodology intends to do. The consequences of one type of error are quite different from those due to another and, as shown in the processing of PEs and TEs through residual analysis (Abur and Expósito, 2004; Monticelli, 1999), such differences will allow to identify the cause of the error. Initially, we will analyze the following facts involving GEs, PEs, and TEs:

- A GE in one measurement will result in a detection of hypothesis test χ^2 (by choosing a $1-\alpha$ probability of false alarm, where α is the confidence level of the test, a numerical value C is obtained via the chi-square distribution table for $\chi^2_{m;\ 1-\alpha}$ such that, in the presence of error, $J_C(\hat{x}) > C$) with a high local CME^N. In other words, the affected measurement will have its CME^N above the chosen error detection threshold β[a].
- A PE in line i–j will spread across all equations in which this parameter is present. Therefore the active and reactive power flows' measurements in that line (on both sides), as well as the active or reactive power injection measurements of the buses i and j will present CME^N above the error detection threshold β;
- A TE can be considered an extreme case of PE, also spreading the error in the vicinity of the affected line (or branch in the bus-branch model). In this case, however, considering that an exclusion-type TE has occurred, i.e., when the line i–j is in operation but does not appear in the topology provided by the network configurator, the active or reactive power flow measurements in that line will not be present in the measurement model. In contrast, the active or reactive power injections in the buses i and j will have very high CME^Ns, usually above 30 standard deviations, due to the power imbalance caused by the transmission line exclusion (Bretas et al., 2017). A similar scenario, with very high CME^Ns, would also occur in the case of an inclusion-type TE, i.e., when the line i–j is not in operation but appear in the topology provided by the network configurator.

[a] This threshold can be chosen according to the desired level for sensitivity to error detection. In the state estimation literature, usually β is equal to three standard deviations of the corresponding measurement, i.e., $\beta=3$ (Abur and Expósito, 2004).

In accordance with the essential characteristics mentioned above, the proposed innovation methodology for detecting, identifying, and correcting GEs in analog measurements, errors in serial and shunt transmission line parameters, and TEs will be described below. In the preliminary tests, the methodology based on the sensitivity analysis of the residuals was followed, however, using the CME^N instead of the r^N. First, the conventional WLS state estimation is performed and then a descending list is constructed with the values obtained for the CME^Ns of all measurements. With this list in hand, it is checked for CME^N of measurements above the β error detection threshold. Subsequently, the error-type identification is performed, applying the algorithm of the proposed innovation methodology for processing GEs, PEs, and TEs, as described below:

1. Read the database of a given measurement set and network configuration.
2. Run WLS state estimation and calculate the estimated state vector ($\hat{x}$), normalized residual vector (r^N), and the projection matrix K.
3. Calculate the innovation vector (II), given by Eq. (7.20), and the error vector composed of measurement in its normalized form (CME^N), given by (7.28).
4. Build a descending list of measurements according to their CME^N values and perform the proposed error detection and identification test using CME^N instead of r^N.
5. Based on the list of Step 4, verify if there are power injection measurements i and j with very high CME^N values (above 15 standard deviations) and no power flow measurements i–j or j–i in the list. If this situation occurs, the i–j line is suspected to be misconfigured, that is, a case of exclusion-type TE has occurred. Then, correct the system topology and return to Step 1. If there are power injection measurements i and j with very high CME^N values and the i–j or j–i power flow measurements are present, then a case of inclusion-type TE has occurred; correct the system topology and return to Step 1. Otherwise, proceed to Step 6.
6. Also from the list in Step 4, determine the i–j lines with power flow measurements in both sides (i–j and j–i) and power injection measurements i and j with CME^N values above the β threshold. If this situation occurs, the i–j branch is suspected of having PEs, so include these suspicious parameters in the state variable vector, perform the joint state/parameter estimation, and return to Step 1. Otherwise, proceed to Step 7.

7. Still based on the list in Step 4, check if there are any measurements CME^N above the error detection threshold. If so, GE is suspected in the measurement vector. Identify the measurement with the largest CME^N and perform the correction through its *CNE* by applying Eq. 7.33. Return to Step 2. If this situation does not occur, proceed to Step 8.
8. End of error processing and state/parameter estimation routine.

Fig. 7.7 illustrates the innovation methodology flowchart developed for error processing.

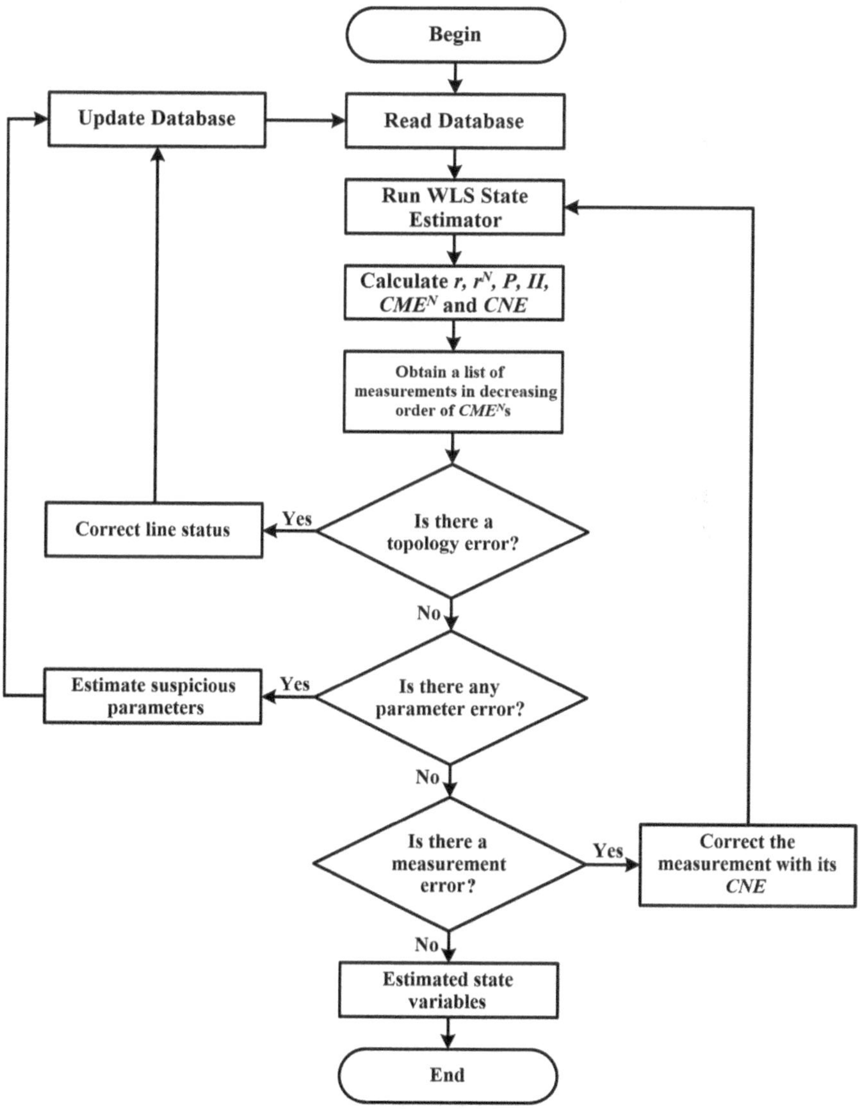

Fig. 7.7 Flowchart of the proposed innovation methodology for processing GEs, PEs, and TEs.

7.6 Examples

Now, examples with different error scenarios will be presented, involving GEs, PEs, and TEs, in order to evaluate the performance of the innovation methodology. Besides that, a list of "problematic" measurements has been built, that is, measurements with Innovation Indexes (*II*) of low magnitude (values below 1.00 or close to that value) which, if contain errors, are difficult to detect by the traditional test of the largest normalized residual (r^N_{MAX}). For the IEEE 14 bus system, considering a measurement plan with 81 measurements (global redundancy equal to 3.0), problematic measurements are highlighted in red (gray in print version) in Tables 7.1 and 7.2.

7.6.1 Example 7.1: Scenario with simultaneous GEs

In this example, GEs were added in the following measurements:

- -6σ in measurement P:12–13;
- -8σ in Q:4;
- 7σ in Q:5–1.

In this case, the test of the largest normalized residual failed, since even though it correctly detected and identified errors in measurements

Table 7.1 IEEE 14—Power injection and voltage magnitude measurements with relatively low *II*.

Power injection			
Active		Reactive	
Measurement	*II*	Measurement	*II*
P:1	4.6892	Q:1	1.4826
P:2	0.2412	Q:3	1.9550
P 3	2.5663	Q:4	0.3460
P:5	0.1720	Q:5	0.2254
P:6	0.6147	Q:6	0.7815
P:7	0.3450	Q:8	1.9666
P:9	1.6814	Q:9	2.0576
P:11	0.5711	Q:11	0.5621
P:13	1.4494	Q:12	0.4783
P:14	1.7947	Q:13	0.9568
Voltage magnitude			
Measurement		***II***	
V:1		9.0319	

Table 7.2 IEEE 14—Power flow measurements with relatively low *II*.

Power Flow			
Active		Reactive	
Measurement	*II*	Measurement	*II*
P:1-2	4.4389	Q:1-2	2.5525
P:1-5	5.0515	Q:1-5	1.1540
P:2-3	3.8946	Q:2-4	1.0722
P:2-5	4.7659	Q:2-5	0.8621
P:3-4	0.7077	Q:3-4	2.1868
P:4-5	2.8285	Q:4-7	2.6940
Q:4-9	3.1560	Q:4-9	4.3655
P:5-6	4.3877	Q:5-6	2.9319
P:6-11	2.1088	Q:6-12	2.1695
P:6-13	2.7912	Q:6-13	2.0516
P:7-9	2.5189	Q:7-8	2.0307
P:9-10	1.1946	Q:9-10	1.0899
P:9-14	1.6614	Q:9-14	1.1926
P:10-11	2.1088	Q:10-11	1.1786
P:12-13	1.0382	Q:13-14	1.2244
P:2-1	4.4773	Q:5-1	0.6652
P:3-2	3.9870	Q:3-2	0.6282
P:4-2	4.3411	Q:4-2	1.7980
P:5-2	4.7894	Q:4-3	2.2702
P:5-4	2.8155	Q:5-4	4.7009
P:7-4	2.5159	Q:7-4	3.2134
P:9-4	3.1560	Q:6-5	1.4796
P:11-6	2.1195	Q:11-6	1.7949
P:12-6	2.1931	Q:12-6	2.0601
P:13-6	2.8150	Q:8-7	1.9666
P:9-7	2.5189	Q:9-7	0.5531
P:10-9	0.2576	Q:10-9	1.0866
P:14-9	1.6829	Q:11-10	1.2061
P:13-12	1.0423	Q:13-12	4.5759
P:14-13	1.1953	Q:14-13	1.1105

P:12–13 and Q:5–1, it failed to detect the error of measurement Q:4, due to its low *II*, as shown in Table 7.3.

Note that the traditional test failed to detect the error in measurement Q:4, which does not occur when the innovation methodology is applied, as shown in Table 7.4.

Table 7.3 IEEE 14—Result of the r_{MAX}^N test in the scenario with GEs.

Step	Wrong measurement	r_{MAX}^N
1	P:12–13	4.5068
2	Q:5–1	3.8392
3	No error detected!	

Table 7.4 IEEE 14—Result of innovation methodology in the GEs.

Original measurement	Measurement with GE
P:12–13 = 0.0161	(P:12–13) − 6σ = 0.0151
Q:4 = 0.0389	(Q:4) − 8σ = 0.0358
Q:5–1 = 0.0214	(Q:5–1) + 7σ = 0.0237

Step 1

$J_C(\hat{x}) = 330.59 > C = \chi^2_{81;0.95} = 103.01 \Rightarrow$ **Error detected!**

Incorrect measurement	*II*	*CME*^{N}	*CNE*
Q:5–1 = 0.0237	0.9594	5.2774	7.6230

Measurement corrected by *CNE*: Q:5–1 = 0.0217 (approx. Error = 1.40%)

Step 2

$J_C(\hat{x}) = 270.62 > C = \chi^2_{81;0.95} = 103.01 \Rightarrow$ **Error detected!**

Incorrect measurement	*II*	*CME*^{N}	*CNE*
P:12–13 = 0.0152	1.0681	−4.3150	−5.9110

Measurement corrected by *CNE*: P:12–13 = 0.0160 (approx. error = 0.62%)

Step 3

$J_C(\hat{x}) = 173.73 > C = \chi^2_{81;0.95} = 103.01 \Rightarrow$ **Error detected!**

Incorrect measurement	*II*	*CME*^{N}	*CNE*
Q:4 = 0.0359	0.4323	−3.3983	−8.5640

Measurement corrected by *CNE*: Q:4 = 0.0390 (approx. error = 0.26%)

Step 4

$J_C(\hat{x}) = 54.36 < C = \chi^2_{81;0.95} = 103.01 \Rightarrow$ No error detected!

The processing of GEs can be followed step-by-step, with corrections of incorrect measurements one at a time, and the reestimation of states. The order followed by the estimator to correct the measurements with GEs follows from the measurement identified with the largest CME^N to the smallest.

By analyzing Table 7.4, it can be noted the effectiveness of the method for treating GEs, even in the case of multiple GEs. Errors between corrected measurement values and the values considered correct presented acceptable values for the state estimation (below 1%), except for measurement Q:5–1, which can be justified by the addition of random noise to the measurement database. However, this error does not affect the state estimation, since it is below the threshold of GE detection.

7.6.2 Example 7.2: Scenario with occurrence of PEs

In this scenario, errors were added in the series and shunt parameters of the following transmission lines:

- −9% on lines 2–3;
- 7% on lines 7–9.

In this PE scenario, the largest CME^Ns were first identified in measurements belonging to lines 2–3, as shown in Table 7.5. Therefore these parameters were added to the state vector for joint state/parameter estimation.

After estimating the parameters of lines 2–3, as shown in Table 7.6, a new state estimation was performed.

Table 7.5 IEEE 14—Result of innovation methodology in the PE scenario.

Step 1			
$J_C(\hat{x})=405.73>C=\chi^2_{81;0.95}=103.01 \Rightarrow$ **Error detected!**			
Descending list of CME^Ns			
Incorrect measurement	***II***	***CME^N***	***CNE***
P:3–2	4.3573	−8.3019	−8.5177
P:2–3	4.2562	8.2889	8.5146
Q:3–2	0.6901	6.1037	10.7464
P:7–9	2.3805	−5.2388	−5.6823
P:9–7	2.3805	5.2388	5.6823
P:2	0.2405	−3.9162	−16.7480

Table 7.6 IEEE 14—Estimated parameters in the PE scenario.

Correction of parameters				
Parameter	**True**	**Incorrect**	**Estimated**	**Approx. error (%)**
g_{2-3}	1.1350	1.0329	1.1350	0.0000
b_{2-3}	−4.7819	−4.3515	−4.7820	0.0021
b_{2-3}^{sh}	0.0219	0.0199	0.0219	0.0000

Table 7.7 IEEE 14—Result of the innovation methodology in the PE scenario.

Step 2			
$J_C(\hat{x})=171.73>C=\chi^2_{81;0.95}=103.01 \Rightarrow$ **Error detected!**			
Descending list of CME^N**s**			
Incorrect measurement	***II***	CME^N	***CNE***
P:9–7	2.3805	5.1523	5.5884
P:7–9	2.3806	−4.6939	−5.0912
Q:9–7	0.5163	3.5003	7.6294
P:9	1.7123	−2.7934	−3.2349
Q:9	2.0910	−2.2121	−2.4520
P:13	1.4510	−1.5026	−1.8249

Table 7.8 IEEE 14—Estimated parameters in PE scenario.

Correction of parameters				
Parameter	**True**	**Incorrect**	**Estimated**	**Approx. error (%)**
g_{7-9}	0.0000	0.0000	0.0000	0.0000
b_{7-9}	−9.0901	−9.7264	−9.0903	0.0022
b_{7-9}^{sh}	0.0000	0.0000	0.0000	0.0000

In this stage, the largest CME^Ns were identified in the measurements of lines 7–9, which can be seen in Table 7.7. The result of processing PEs, after estimation of erroneous parameters, is shown in Table 7.8. After reestimating with the corrected parameters, it was obtained $J_C(\hat{x}) = 53.98 < C = \chi^2_{81;0.95} = 103.01$, so no error was detected.

7.6.3 Example 7.3: Scenario with simultaneous GE, PE, and TE

In this last scenario, the most serious case of errors will be simulated, where GE, PE, and an exclusion TE will occur simultaneously in the power system. These are described below:

- 6σ GE in measurement Q:2 = 0.3084;
- PE of 5% in the parameters of lines 2–5;
- Exclusion-type TE from lines 6 to 13.

In the first step of error processing, a decreasing list of CME^Ns was obtained as shown in Table 7.9. By analyzing it, one can note the extremely high values for the CME^Ns of the injection measurements in buses 6 and 13. One can also notice that the absence of power flow measurements in this branch, which characterizes an exclusion-type TE. Consequently, the line status is corrected in the database and a new estimation is performed.

In the second step of error processing, the decreasing list of CME^Ns was obtained as shown in Table 7.10. By analyzing it, we have several measurements from lines 2 to 5 with the CME^Ns above the error detection threshold, characterizing a case of PE. Therefore the parameters of this line are added to the state vector to perform the estimation of states and parameters. After the joint estimation, the parameters of lines 2–5 are corrected in the database as described in Table 7.11 and a new estimation of state is done.

In the third step of error processing, the decreasing list of CME^Ns was obtained shown in Table 7.12. By analyzing it, one can notice that only the Q:2 measurement presented the CME^N above the error detection threshold, so the GE was identified and this measurement was corrected

Table 7.9 IEEE 14—Results obtained in the first step: simultaneous GE, PE, and TE scenario.

Step 1			
$J_C(\hat{x}) = 5.52 \times 10^4 > C = \chi^2_{81;0.95} = 103.01 \Rightarrow$ **Error detected!**			
Descending list of CME^Ns			
Incorrect measurement	***II***	***CME^N***	***CNE***
P:13	3.8595	−126.4163	−130.5907
P:6	0.9248	110.5415	162.8071
Q:13	2.5864	−104.2338	−111.7535
Q:6	1.1408	102.5529	136.3759
Q:5	0.2335	61.3768	169.9359
Q:12	0.4278	−37.4456	−95.2101

Table 7.10 IEEE 14—Results obtained in the second step: simultaneous GE, PE, and TE scenario.

Step 2			
$J_C(\hat{x})=162.26>C=\chi^2_{81;0.95}=103.01 \Rightarrow$ **Error detected!**			
Descending list of CME^N**s**			
Incorrect measurement	*II*	CME^N	*CNE*
Q:2	3.7594	6.8191	7.0562
P:5–2	6.0398	5.1116	5.1812
P:2–5	6.0007	−4.9507	−5.0190
Q:5–2	1.8982	3.9892	4.3787
Q:5	0.2279	−3.4619	−11.0819
Q:2–5	0.9762	−3.3149	−3.3139

Table 7.11 IEEE 14—Estimated parameters in the simultaneous GE, PE, and TE scenario.

Correction of parameters				
Parameter	**True**	**Incorrect**	**Estimated**	**Approx. error (%)**
g_{2-5}	1.7011	1.7862	1.7050	0.2292
b_{2-5}	−5.1939	−5.4536	−5.2002	0.1213
b^{sh}_{2-5}	0.0173	0.0182	0.0173	0.0000

Table 7.12 IEEE 14—Results obtained in the third step: simultaneous GE, PE, and TE scenario.

Step 3			
$J_C(\hat{x})=110.06>C=\chi^2_{81;0.95}=103.01 \Rightarrow$ **Error detected!**			
Descending list of CME^N**s**			
Incorrect measurement	*II*	CME^N	*CNE*
Q:2	3.7275	5.8862	6.0943
Q:1	1.7083	1.6424	1.9031
Q:5–1	0.8945	−1.3911	−2.0866
Q:5–2	1.9733	1.2657	1.4189
P:11	0.6727	1.2079	2.1642
Q:9–7	2.9063	−1.0804	−1.1426
Measurement with $\|CME^N\|\geq 3$	*II*	CME^N	*CNE*
Q:2 = 0.3282	3.7275	5.8862	6.0943
Measurement corrected by *CNE*: Q:2 = 0.3094 (approx. error = 0.3242%)			

through his *CNE*, obtaining the value for Q:2 = 0.3094, very close to the original value of the database Q:2 = 0.3084 (approximation error = 0.3242%) as shown in Table 7.12. After correcting the measurement, a new estimation was performed and no error was detected, since the result was $J_C(\hat{x}) = 22.39 < C = \chi^2_{81;0.95} = 103.01$.

Note 7.2: This scenario was simulated in order to show the robustness of the innovation methodology, both in relation to the simultaneous occurrence of the three types of errors, and to the fact that GE occurs in an incident manner in the vicinity of the line with PE. Despite this unfavorable scenario, the methodology was able to identify the GE after correcting the wrong parameters.

References

Abur, A., Expósito, A., 2004. Power Systems State Estimation: Theory and Implementation. Marcel Dekker, Inc., New York, USA.

Allan, M., Laughton, M., 1973. A General Algorithm for Estimating Power System Variables and Network Parameters. IEEE PES Summer Meeting, Anaheim, USA, pp. 1–6.

Alsaç, O., Vempati, N., Stott, B., Monticelli, A., 1998. Generalized state estimation. IEEE Trans. Power Syst. 13 (3), 1069–1075.

Bretas, N.G., London Jr., J.B.A., 2010. Recovering of Masked Errors in Power Systems State Estimation and Measurement Gross Error Detection and Identification Proposition. In: IEEE PES General Meeting, Minneapolis, USA, pp. 1–6.

Bretas, N.G., Piereti, S.A.R., 2010. The Innovation Concept in Bad Data Analysis Using the Composed Measurements Errors for Power System State Estimation. IEEE PES General Meeting, Minneapolis, USA, pp. 1–6.

Bretas, N.G., London Jr., J.B.A., Alberto, L.F.C., Benedito, R.A.S., 2009. Geometrical Approaches for Gross Errors Analysis in Power Systems State Estimation. IEEE Powertech Conference, Bucharest, Romania, pp. 1–7.

Bretas, N.G., Bretas, A.S., Piereti, S.A.R., 2011. Innovation concept for measurement gross error detection and identification in power system state estimation. IET Gen. Trans. Distrib. 5 (6), 603–608.

Bretas, N.G., Piereti, S.A.R., Bretas, A.S., Martins, A.C.P., 2013. A geometrical view for multiple gross errors detection, identification, and correction in power system state estimation. IEEE Trans. Power Syst. 28 (3), 2128–2135.

Bretas, A.S., Bretas, N.G., Carvalho, B.E.B., Baeyens, E., Khargonekar, P., 2017. Smart grids cyber-physical security as a malicious data attack: an innovation approach. Electr. Power Syst. Res. 149, 210–219.

Debs, A., 1974. Estimation of steady-state power system model parameters. IEEE Trans. Power App. Syst. 93 (5), 1260–1268.

Monticelli, A., 1999. State Estimation in Electric Power Systems: a generalized approach. Kluwer Academic Publishers, Boston, USA.

Simões Costa, A.J., Leão, J.A., 1993. Identification of topology errors in power system state estimation. IEEE Trans. Power Syst. 8 (4), 1531–1538.

Slutsker, I., Mokhtari, S., Clements, K., 1996. Real time recursive parameter estimation in energy management systems. IEEE Trans. Power Syst. 11 (3), 1393–1399.

van Cutsem, T., Quintana, V., 1988. Network parameter estimation using online data with application to transformer tap position estimation. IET Gen. Trans. Distrib. 135 (1), 31–40.

Zarco, P., Expósito, A., 2000. Power system parameter estimation: a survey. IEEE Trans. Power Syst. 15 (1), 216–222.

CHAPTER 8

Dynamic state estimation in electric power systems

The previous chapters of this book focus on the static state estimation problem. This chapter addresses the dynamic state estimation problem. As mentioned in Chapter 1, in the case of the dynamic estimator[a], the variations of the quantities of interest, relative to the variable "time", are considered in the network modeling. On the other hand, in the static estimator, the network model considered is not time-dependent, translating into a "photograph" of the system at a fixed time-instant.

Note 8.1: The authors would like to acknowledge the PhD student Júlio Augusto Druzina Massignan, who provided useful comments about this chapter and was responsible by the elaboration of all the simulation results presented in this chapter.

8.1 Introduction

The word estimation can be applied to describe three keywords: prediction, filtering, and smoothing. Filtering is popularly a well-known word, but the same cannot be said of the words prediction and smoothing. Almost everyone is aware of attempts to remove impurities from water by filtering since the beginning of civilization. In communication, people always receive information accompanied by noise. The information originally transmitted has to be recovered and so must be filtered. On the other hand, the impurities must be removed.

A good way for understanding the meaning of the word smoothing is to understand the action of the human brain when it tackles the problem of reading hastily written handwriting. Each word is tackled sequentially and when a word is encountered, which is particularly difficult to interpret,

[a] A broad literature review on dynamic estimators can be found in Shivakumar and Jain (2008); Coutto Filho and Souza (2009) and Zhao et al. (2019).

Cyber-Physical Power Systems State Estimation
https://doi.org/10.1016/B978-0-323-90033-1.00004-4

several words after the difficult word, as well those before, may be used in an attempt to deduce it.

The human brain can be used again to exemplify the prediction situation. When attempting to catch a ball, the future trajectory of the ball has to be predicted in order to position a catching hand correctly. This task becomes more difficult when the ball is subject to random disturbances such as wind.

In terms of estimating the state variables of a system, one has a set of measurements of the process up to the present time and wants to know, with some confidence, what the system will be in the future. This is a prediction problem. If it is necessary to know what the system was in the past, then it is a smoothing problem; and if it is necessary to know what the system is at present, then it is a filtering problem.

As previously stated, estimation may be applied in these three concepts. In this book, it will be used only as it is applied to filtering.

8.2 The history of estimation

State estimation has existed for a long time. People have been trying to interpret observations since the beginning of civilization. The study by Neugebauer (1952) showed that the Babylonians used a rudimentary form of the Fourier series for estimation purposes. Attempts to minimize various functions of errors, which are a kind of beginning of the estimation theory, can be attributed to Galileo Galilei, back in 1632, as stated by Wilson (1972). Many well-known authorities including Euler, Lagrange, Laplace, Bernoulli, and others attempted, in one way or another, to make estimations.

The first to use the method of least squares was Gauss (1795), although it was not published until 1805 by Legendre (1805). Many extensions of Gauss approach have been proposed, but one which brings something knew was proposed recently by Bretas and Bretas (2018). The authors (Kolmogorov, 1940; Krein, 1945; Wiener, 1964) made the first studies of least squares estimation in stochastic process. At that time, Kolmogorov (1940) gives a comprehensive treatment of the prediction problem for the discrete time stationary process. Krein (1945) extended the work of Kolmogorov to continuous time by use of a bilinear transformation. However, none of these researchers paid attention to explicit formulas for the optimum predictor itself, which is vital in many other applications such as antiaircraft fire control.

The author (Wiener, 1964), independently, formulated the continuous-time linear prediction problem and derived an explicit formula for the optimum predictor. He also considered the estimation problem when the process is corrupted by noise.

In the 1960s, Kalman (1960) and Kalman and Bucy (1961) made an important contribution to space age electronics. He developed an algorithm that seemed particularly proper to the dynamic state estimation problem, which was brought forward by the advent of the space age. People in this area took Kalman's idea and developed it into programs that were used successfully in many experiments with space. Kalman replaced the conventional specification of the filtering problem in terms of signal and noise covariance functions with one in which state-space models were specified for the signal and the noise. Kalman's main contribution was the proof that under the condition of controllability and observability, the optimal filter is stable in the sense that effects of errors and round off and others, computational errors will die out asymptotically.

8.3 Dynamic state estimation modeling

As far as we understand, dynamic state estimation in electric power systems means estimating the network states, voltages magnitude, and phases, for an interconnected power system, considering the time relation between states, and can be modeled as

$$\underline{x}_{k+1} = F_k\underline{x}_k + G_k\underline{w}_k + \Gamma_k\underline{u}_k$$
$$\underline{z}_k = H_k\underline{x}_k + \underline{v}_k \tag{8.1}$$

where $\underline{x}_k$ is the state vector at time k, $\underline{w}_k$ is a known input, and $\underline{u}_k$ is the state noise.

Equivalently, $\underline{z}_k$ is the observation vector and $\underline{v}_k$ is the observation, or measurement, noise. F_k is the state transition matrix, G_k is the matrix relating the states to the input $\underline{w}_k$, and H_k is the matrix relating the measurements to the state at time k, all them with the appropriate dimensions, as presented in Fig. 8.1.

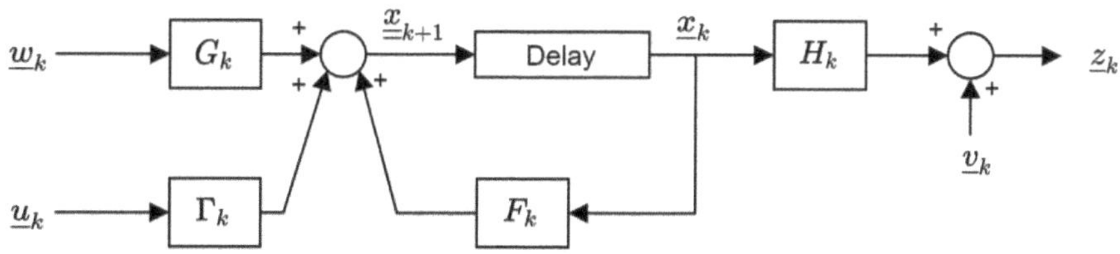

Fig. 8.1 Signal model with external known input.

There are many dynamic state estimation (DSE) formulations, such as the Kalman filter (KF) formulation—it's like a family of formulations. We are going to use the KF family formulation to present the DSE development.

8.3.1 Statement of the discrete KF problem

For the linear, finite-dimensional, discrete system described by the previous equations, with $\underline{v}_k$ and $\underline{w}_k$ being independent, Gaussian white noise processes with $E\left[\underline{v}_k \underline{v}_j^T\right] = \delta_{kj} R_{kj}$ and $E\left[\underline{w}_k\, \underline{w}_j^T\right] = \delta_{kj} Q_{kj}$ and u_k a known input.

Suppose $\underline{x}_0$, the initial value of the system state variables, being a Gaussian random variable. Assume they have mean $\overline{\underline{x}}_0$ and covariance matrix $\sum_{0|-1}$, the KF solution consists of estimating $\underline{x}_{k|k-1}$ and $\underline{x}_{k|k}$ and their associated error covariance matrices, $\sum_{k|k-1}$ and $\sum_{k|k}$. If the assumption of a Gaussian distribution for $\overline{\underline{x}}_0$, $\underline{v}_k$, and $\underline{w}_k$ is dropped, but still remains described by their first- and second-order moments, all the conclusions and formulas are precisely the same.

8.3.2 Solution of the KF problem

In this section, the solution of the problem will be stated. The proofs will not be presented in the book, although they are quite simple to prove. The solution is composed of two steps: the prediction step and the estimation step.

(1) Solution of the prediction step:

The KF prediction step is given by.

$$\hat{\underline{x}}_{k+1|k} = F_k \hat{\underline{x}}_{k|k-1} + K_k\left[\underline{z}_k - H_k \hat{\underline{x}}_{k|k-1}\right] + \Gamma_k \underline{u}_k \tag{8.2}$$

With $\hat{\underline{x}}_{0|-1} = \overline{\underline{x}}_0$. The Kalman gain is given by

$$K_k = F_k \sum\nolimits_{k|k-1} H_k^T \left[H_k \sum\nolimits_{k|k-1} H_k^T + R_k\right]^{-1}, \tag{8.3}$$

if the inverse exists. The inverse is guaranteed with R_k being a positive definite matrix.

The conditional error covariance matrix $\sum_{k|k-1}$ is computed recursively by the known Riccati equation:

$$\sum\nolimits_{k+1|k} = F_k\left[\sum\nolimits_{k|k-1} - \sum\nolimits_{k|k-1} H_k^T \left(H_k \sum\nolimits_{k|k-1} H_k^T + R_k\right)^{-1} H_k \sum\nolimits_{k|k-1}\right] F_k^T + G_k Q_k G_k^T \tag{8.4}$$

(2) Solution of the estimated step:

The KF estimating solution is given by

$$\hat{\underline{x}}_{k|k} = \hat{\underline{x}}_{k|k-1} + \sum\nolimits_{k|k-1} H_k^t \left(H_k \sum\nolimits_{k|k-1} H_k^t + R_k \right)^{-1} \left(z_k - H_k \hat{\underline{x}}_{k|k-1} \right), \quad (8.5)$$

assuming that the inverse exists. Again, this is guaranteed with R_k, a positive definite matrix. The associated error covariance matrix is given by

$$\sum\nolimits_{k|k} = \sum\nolimits_{k|k-1} - \sum\nolimits_{k|k-1} H_k^T \left(H_k \sum\nolimits_{k|k-1} H_k^T + R_k \right)^{-1} H_k \sum\nolimits_{k|k-1} \quad (8.6)$$

Observations:

(1) The inputs of the filter are the measurements $\mathbf{z}_k$ and the known bias u_k, if it exists.

(2) Suppose the signal model under consideration is time invariant, and the input and output noise processes are stationary, then the KF will normally be time varying despite time invariance and stationarity of the signal model. This is because different sets of measurements are received every time step.

8.4 Dynamic state estimation GE analysis

In DSE, instead of measurement errors only, one has anomalies, which contains also large load changes. In this case, the state transition matrix may not reflect the hull transition of states. That shows the difficulty in modeling the system for DSE purposes.

The state modeling fluctuation may be represented by a white Gaussian noise with covariance Q_k. This technique may incur in some difficulties: first if it is assumed a high covariance so that the big load change is inside the modeling, the estimate may be significantly poor for the case of small load fluctuations. In contrary, if that noise covariance is small, in order to represent well the small load fluctuations, the model will not simulate large load fluctuations, but, in this case, it could be used to identify the anomaly for large load changes (Kalman and Bucy, 1961).

We propose to model the state transition matrix as an identity matrix with small covariance matrix, so that it will represent a stationary system behavior, and possible change in the stationarity load level can be online detected. One should be aware that, sometimes, measurement GE could leave the impression of stationarity level change too.

Suddenly large load changes, as well as measurement GEs, will be named here as "anomaly". However, they are anomalies of quite different

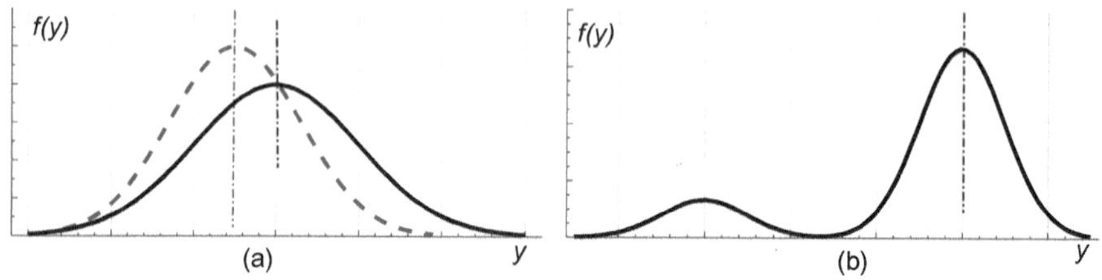

Fig. 8.2 Distribution function of the new information: (A) *dashed line* for normal distribution and *solid line* for sudden load change situation, and (B) measurement GE case.

characteristics and should be, in this way, differentiated. One should be aware that in GE situation there exist equations "inconsistencies" or "a structural discontinuity" of the equations describing the DSE process. In case of large load change, on the other hand, there is no structural discontinuity in system equations (Bretas, 1986). In this book, the differentiation between these anomalies is going to be performed through the analysis of the innovation process distribution.

The innovation for the measurement GEs situation will present a large asymmetry characteristic due to the structural discontinuity. In the case of stationarity of the load level change, otherwise, the symmetry related to the innovation average value will keep its value, but dislocated from zero, since the innovation, for this case, will not have normal distribution anymore. These characteristics are depicted in Fig. 8.2.

8.5 Proposed dynamic state estimation formulation for GE analysis

The power system state is uniquely defined by an n-dimensional $\underline{x}$ vector consisting of the voltages magnitudes and their phase angles, except one of them, taken as reference. The system state will be modeled as

$$\underline{x}_{k+1} = \underline{x}_k + \underline{w}_k, \tag{8.7}$$

with $\underline{w}_k$ being a white Gaussian, noise vector with $E\left[\underline{w}_k \, \underline{w}_j\right] = \delta_{kj} Q_{kj}$ and $Q = \alpha^2 \, diag\left(q_i^2\right) = \alpha^2 \Gamma$.

The previous q_i is the maximum ratio of the state i obtained from historical data of bus i. The measurement equations are given by

$$\underline{z}_k = h(\underline{x}_k) + \underline{v}_k \tag{8.8}$$

and $E\left[\underline{v}_k \, v_j^T\right] = \left[\delta_{kj} R_{kj}\right]$, $E[\underline{v}_k] = \underline{0}$.

The estimate $\underline{\hat{x}}$ of $\underline{x}$ is obtained using the KF theory:

$$
\begin{aligned}
\sum\nolimits_{k+1|k} &= \sum\nolimits_{k|k} + Q_k \\
K_k &= \sum\nolimits_{k|k} H_k^T \left[H_k \sum\nolimits_{k|k-1} H_k^T + R_k \right)^{-1} \\
\underline{\hat{x}}_{k|k} &= \underline{\hat{x}}_{k|k-1} + K_k \left[z_k - \underline{h}\left(\underline{\hat{x}}_{k|k-1} \right) \right] \\
\sum\nolimits_{k|k} &= [I - G_k H_k] \sum\nolimits_{k|k-1},
\end{aligned}
\tag{8.9}
$$

with H_k the Jacobian matrix of $\underline{h}$ computed at $\underline{x} = \underline{\hat{x}}$.

The hull differentiation between the anomalies, GE and sudden load change, will be based on the deformation of the innovation distribution function (Bretas, 1989).

In what follows, the innovation process is introduced in order to better differentiate between the anomalies. One should be aware that in using this approach one is assuming the system behavior as stationary by parts. With that approach, detecting the changes of stationarity load level becomes a crucial point.

8.5.1 The innovation in the DSE process

The system innovation is modeled by

$$\underline{v}_k = \underline{z}_k - \underline{h}\left(\underline{\hat{x}}_{k|k-1} \right). \tag{8.10}$$

As one knows, $\underline{v}_k$ is a white Gaussian process, so that $E = [\underline{v}_k] = \underline{0}$ and $E = \left[\underline{v}_k v_j^T \right] = H_k \sum_{k|k-1} H_k^T + R_k$.

Normalizing the innovation process, one obtains

$$
\begin{aligned}
\lambda_{k,i} &= \lambda_{k,i} / \rho_{k,i} \\
\rho_{k,i}^2 &= H_k \sum\nolimits_{k|k-1} H_k^T + r_{k,i}^2
\end{aligned}
\tag{8.11}
$$

where $v_{k,i}$ is the ith component of $v_k$and $H_{k,i}$ is the ith line of H_k. In this way, the random variable Λ_k whose sample is the normalized innovation process presents a normal distribution with average zero and unitary covariance. One should be aware that in case of GE situation, that would not be in fact an innovation.

All the differentiation between the anomalies, GEs and changes of stationarity load level, are based on the innovation distribution deformation, as previously described.

8.6 Anomaly detection

As previously stated, two kinds of anomaly are considered, that is, the abrupt change of system states and measurement GEs. One should be aware that the measurement value z_k, in case of gross measurement GE, and the predicted state value will present significant errors in the innovation equation.

When these not usual situations occur, the innovation distribution (Λ) in the normalized form is considerably different of the Λ distribution for the case of stationarity as illustrate in Fig. 3.1, where $f(\lambda)$ is the probability distribution function for λ. It is not difficult to understand that for measurement GE case, the normalized innovation corresponding to the measurements with GE is affected, and if those measurements are deleted or corrected, the distribution for Λ is equivalent to the stationary case situation. On other hand, for the abrupt state changes case, a significant number of elements of the innovation process are affected, and the Gaussianity behavior will stay valid, but with an average magnitude not zero, that is, the symmetry related to the nonzero average value will stay valid, which does not occur for the measurement GE situation.

The anomaly will be detected by the test:

$$|\lambda_{k,i}| < a, i = 1, 2, \ldots, n \tag{8.12}$$

with a being a chosen threshold value. If at least one element of the innovation process is out of the prespecified region, a situation of anomaly is characterized.

8.6.1 Anomaly discrimination

The anomaly discrimination will be performed through the deformation or deviation of the new information distribution. This index of deformation is made through the asymmetry index of the innovation distribution related to the average:

$$\gamma_k = {}^{M_{3,k}}\!/_{\sigma_k^3}, \tag{8.13}$$

with $M_{3,k}$ the third moment of the distribution, at time k, and σ_k the standard deviation of that distribution. The third moment is given by

$$M_{3,k} = E\left[\Lambda_k^3\right] - 3\mu_k E\left[\Lambda_k^2\right] + 2\mu_k^3, \tag{8.14}$$

with

$$\sigma_k^3 = E\left[\Lambda_k^2\right] - \mu_k^2, \mu_k = E[\Lambda_k] \tag{8.15}$$

The deformation tends to zero when the distribution tends to symmetric and so the constant γ_k tends to zero, in the normal distribution. Therefore for measurement GE, the deformation, then γ_k becomes large because the symmetry is lost. By other side, for abrupt change of the state vector, the symmetry is not lost and γ_k will have a low value.

As a conclusion, calculating γ_k for the normalized innovation one will have

(a) $\gamma_k > |\gamma_{max}| \rightarrow$ Measurement GE situation.
(b) $\gamma_k < |\gamma_{max}| \rightarrow$ Sudden state variable change.
where $|\gamma_{max}|$ is obtained by experience and at same time by simulations.

8.6.2 Discussion of possible advantages or disadvantages between static and dynamic state estimation

(a) Modeling

The static state estimation modeling is extremely easy; it corresponds to a photo of the system in an instant of time.

The dynamic state estimation modeling, on the other hand, is not easy. Since, in this case, the state variables in different time instants are required, the state transition matrix is required, and that is, until today, not a well-solved problem.

(b) Numerical solution

The static state estimation brings the advantages of keeping the Y_{bus} sparsity characteristics in the iterative process of solution.

That sparsity characteristic is not kept in the dynamic state estimation case. The reason of that is the addition of the matrix in the iterative process.

8.7 Presentation of performed tests and analysis of results

8.7.1 Introduction

The main objective of this section is to describe the performed tests and analyze the estimation for the proposed anomaly detection model. The algorithm behavior for anomalies differentiation, through the asymmetry index of the distribution curve of the innovation, is presented. In the DSE model, to be used at this time, no bias is going to be incorporated.

8.7.2 Case 1—Dynamic state estimation anomaly and GE detection

Case summary: Estimation corresponding to 100s of data, from different time instant, obtained in a sequence, with introduction of measurement GE.

The data to perform the GE analysis will be the same used for the case of anomaly detection case. The IEEE 14 bus transmission test system is used for the tests in this case. The measurements are shown on the power system diagram in Fig. 8.3.

The data were obtained from the standard load flow scenario available at Christie (1993). A stationary load condition is considered under the first 20 s of the simulation, that is, an ideal situation where the loads are kept the same fixed load values. A load increase at a constant rate of 0.2% per second follows from $t=20$s to $t=60$s, that is, a constant load ramp for all nodes, followed by more 10 s of stationarity condition. Then, a sudden load increase of 5.0% occurs from $t=70$s to $t=90$s at four specific nodes (nodes 3, 4, 13, and 14). Fig. 8.4 illustrates the total net load and the total generation of the IEEE 14 bus system during this sequence of events.

The simulation consists of estimation corresponding to 100 s of data, from different time instants, obtained in a sequence. The measurements were obtained with a time interval of 1 s from each other, with the introduction of measurement GEs, but also with random error added to the measurements.

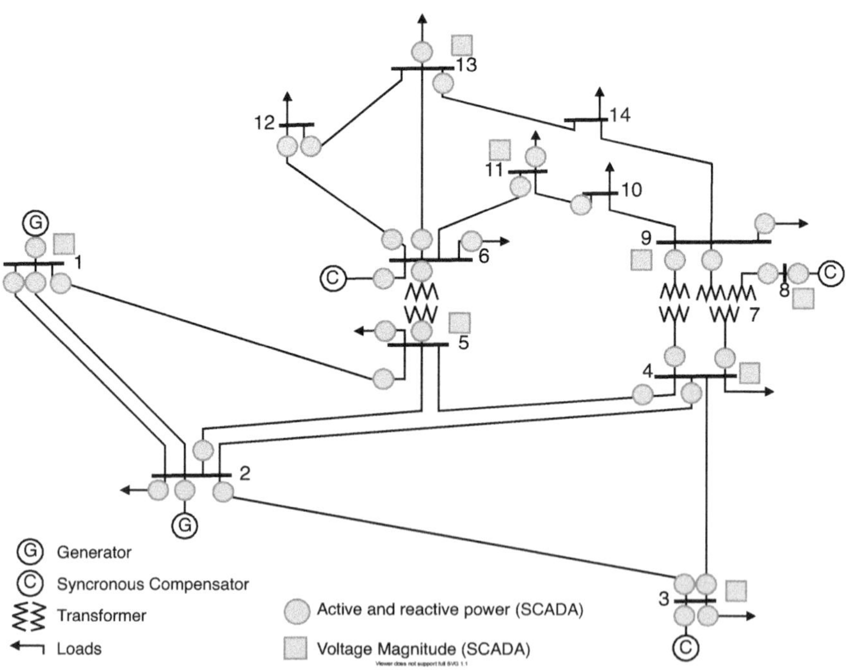

Fig. 8.3 The IEEE 14 bus transmission system and the associated measurements considered in the test.

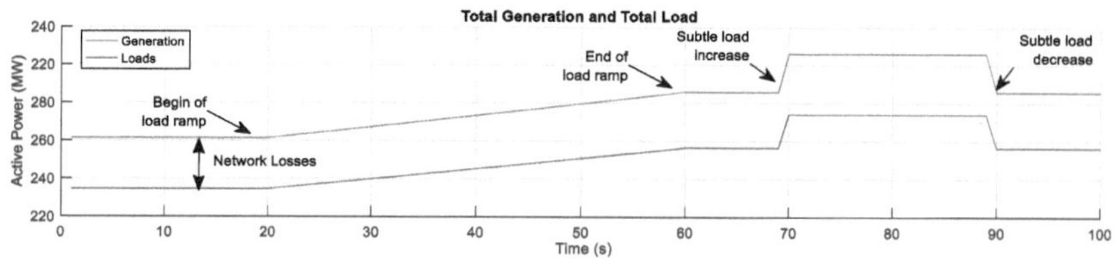

Fig. 8.4 Total net load and generation in the IEEE 14 bus test system for the time period of the simulation.

The measurement values were obtained by including random Gaussian error of 2% for active and reactive power measurements and 1% for voltage magnitude measurement.

A GE was inserted at the active and reactive power flow measurement from nodes 3 to 4 (Pkm_3–4 and Qkm_3–4), from $t=37$s to $t=42$s. It was added an error of 30 times the respective measurement standard deviation.

1. GE is introduced at measurement Pkm_3–4 and Qkm_3–4 from $t=37$s to $t=42$s.

Regarding the dynamic state estimator, the process noise covariance matrix (Q) was obtained considering a 0.03% of possible random variation relative to the previous values of the state variables, that is, a diagonal matrix where each element is given as 0.03% from the latest estimated values. The thresholds for anomaly detection and GE discrimination were assumed as 3.0 and 2.0, respectively.

8.7.2.1 Analysis of results

Fig. 8.5 illustrates the estimated state variables in comparison to the reference load flows for node 14 of the system. Fig. 8.6 presents the estimation error boxplot between the estimated state variables and the reference values, obtained in the load flow calculation for all nodes. Both figures illustrate the KF performance deteriorates after anomalies occur in the system, especially for the phase angle. This is expected, since, during such events, the assumption that the state transition matrix is equal to an identity matrix becomes false. On the other hand, that the previous state variables are good approximations for the latest ones is far from reality. It is noteworthy that any special tuning has been performed in this example, in order to illustrate the negative effect of unknown anomalies into the estimation process. It motivates the application of advanced techniques for the detection of such

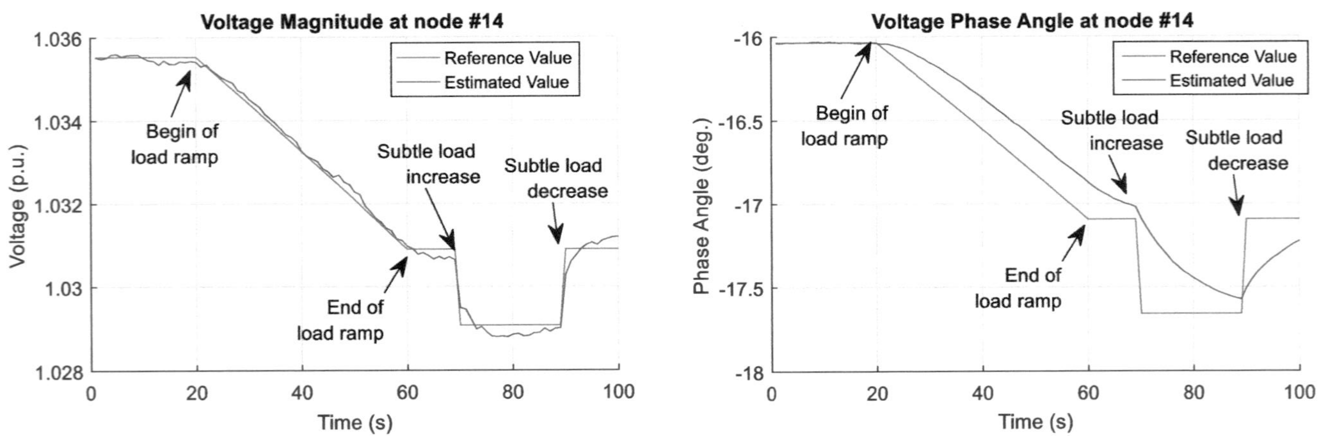

Fig. 8.5 Example of estimated state variables (voltage magnitude and phase angle) at bus 14 during the simulations.

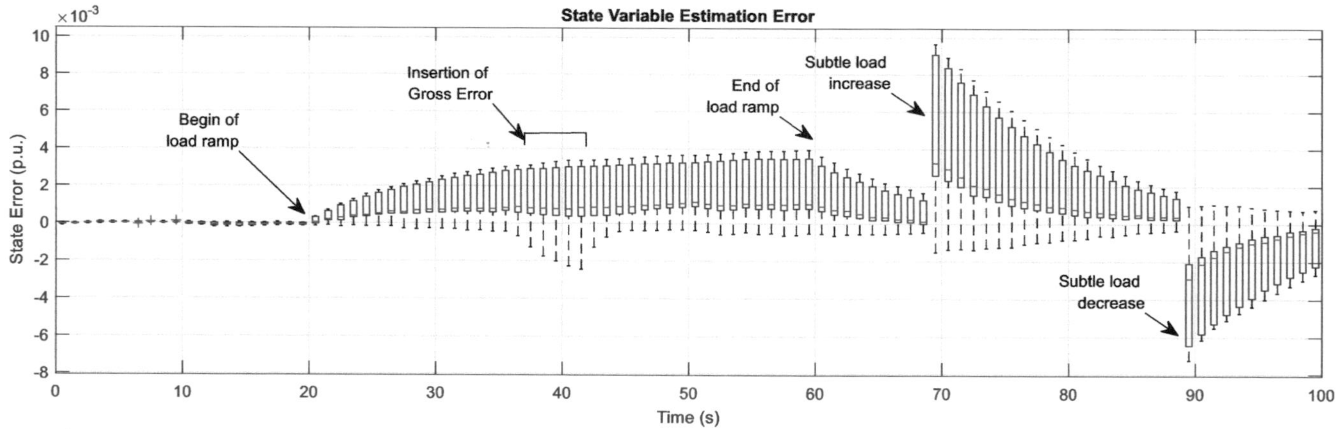

Fig. 8.6 Boxplot of estimation error in all state variables at each instant of the simulation, using the Kalman filter approach. The estimation error was obtained from the difference between the load flow reference values and the final estimated state variables.

anomalies, which will be shown in the next figures, using the method described in the previous sections.

In order to detect such anomalies and to discriminate them from GEs, the normalized innovation index, along with the asymmetry index, was calculated. Fig. 8.7 illustrates the maximum normalized innovation index from all measurements in each instant of the simulation and the detection threshold, which comprise the first step of the method to detect anomalies and GEs. Fig. 8.8 illustrates the calculated asymmetry index in each instant of the simulation and the detection threshold, which comprise the second step of the method to discriminate GEs.

It is clearly shown that, just after the inclusion of GE in the measurement, the asymmetry index increases to 4.354, at the estimation step no. 37. In the other entire estimation steps, that index is lower than the threshold, in the order of 0.5. Consequently, the conclusion is for the occurrence of GE. Looking for the normalized innovation, clearly the measurement with error is the active power measurement Pkm_3–4, with normalized innovation index = 24.244. The second largest is from the reactive power measurement Qkm_3–4, also contaminated with GE, with normalized innovation index = 14.243. Many other tests were performed, but with GE in other measurements, all them detecting and identifying the measurement with error, as can be seen on the tables.

Besides the proper detection of and discrimination of a GE, the anomaly detection through the asymmetry index properly assessed the occurrence of events on the system, from the start of load ramp as well as the immediate detection of sudden load changes. For instance, during the sudden load increase of 5%, at the estimation step no. 70, the maximum normalized innovation becomes 15.870, while the asymmetry index is equal to 1.149, below the threshold 2.0. Therefore the conclusion is for the occurrence of an anomaly in the system dynamics, and not the presence of GEs. This illustrates the effect of changes on the state variables that are not adequately encompassed by the dynamic model. Once again, the assumption that the model is always correct does not hold on practical applications. Thereby, it is essential the association of dynamic state estimators with advanced anomaly detection methods. Other tests were performed, with different level of events on the system, all of them discriminating anomalies from GE, as can be seen on the tables.

The detailed results for the normalized innovation, the final estimated state variables in comparison to the reference load flow values, and the asymmetry index at some instants are shown in the next tables. The values with

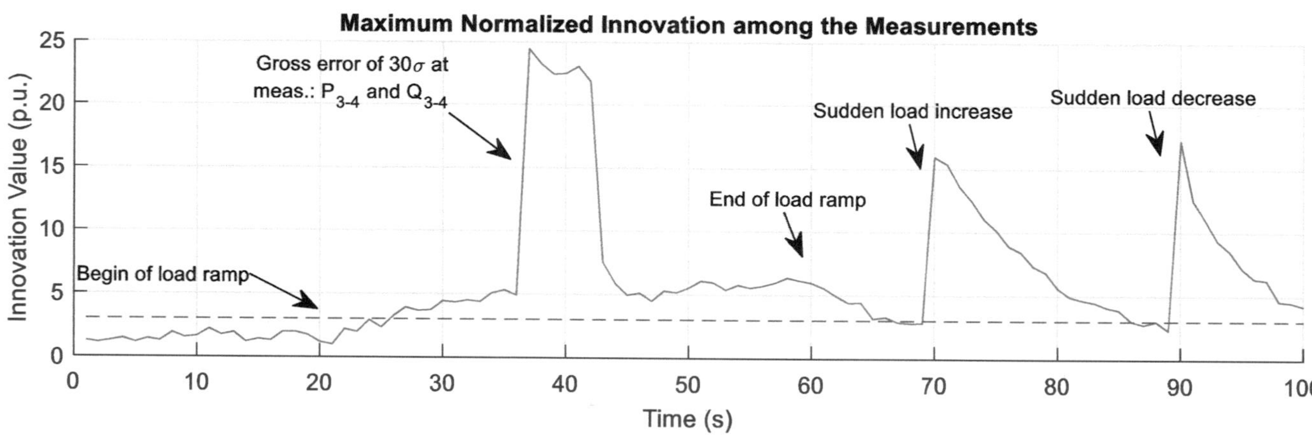

Fig. 8.7 Maximum normalized innovation index in the IEEE 14 bus test system during the simulations.

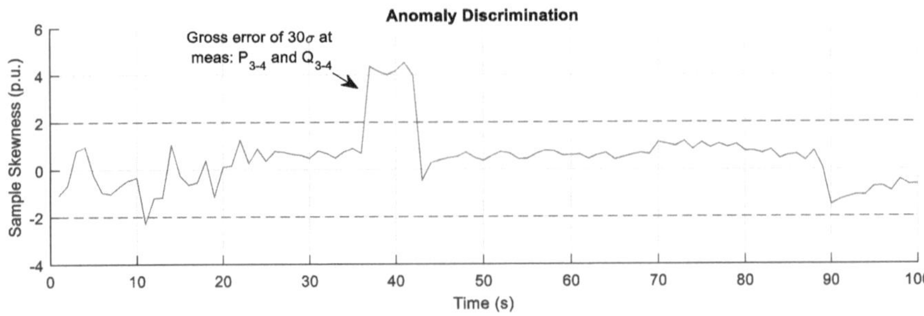

Fig. 8.8 Asymmetry index from the sample skewness in the IEEE 14 bus test system during the simulations.

largest absolute normalized innovation are highlighted in bold (largest value and second largest). On the table, one has

Column 1: measurement variation ("innovation");

Column 2: normalized covariance of the innovation;

Column 3: Reference values (voltage magnitude and voltage phase angle) obtained from the load flow calculation;

Column 4: Estimated state values—innovation asymmetry index.

The instants and respective tables are the following:

1. Instant $t=10$s: initial stationary condition with fixed load values (Table 8.1);
2. Instant $t=30s$: in the presence of a load ramp of 0.2% in all loads (Table 8.2);
3. Instant $t=40$s: in the presence of a load ramp of 0.2% in all loads and with the inclusion of a GE of 30 standard deviation at the active and reactive power flow measurement Pkm_3–4 and Qkm_3–4 (Table 8.3);
4. Instant $t=70$s: in the presence of a sudden load increase of 5% in four nodes (Table 8.4);
5. Instant $t=90$s: in the presence of a sudden load decrease of 5% in four nodes (Table 8.5).

As can be seen in Table 8.1, the normalized innovation keeps its absolute value below the detection threshold of 3.0, evidence that any anomaly has occurred in the system during this period. Besides, the asymmetry index also keeps its absolute value below the threshold of 2.0. It is another evidence for stationarity and the absence of any anomalies. This is in accordance with the assumed stationary condition for the first 20 s of the simulation.

The normalized innovation above the detection threshold, for measurements *Qkm_2–6* and *Qmk_5–3*, indicates the presence of an anomaly at this step.

Table 8.1 Estimated step no. 10: initial stationary condition.

Measurement	Normalized innovation	Measurement	Normalized innovation	State variable	Reference value	Estimated value
P_1	0.0137286	Pmk_7–6	0.1674940	V_1	1.060000	1.060003
P_2	−0.1657214	Pmk_12–7	0.3844746	V_2	1.045000	1.045001
P_3	−0.2319808	Pmk_13–7	−0.2832286	V_3	1.010000	1.009982
P_5	0.0127150	Pmk_9–8	−0.0000014	V_4	1.017671	1.017686
P_6	0.0622195	Pmk_10–8	−0.0343918	V_5	1.019514	1.019524
P_8	−0.0000014	Qkm_2–3	−0.2027471	V_6	1.070000	1.070011
P_9	−0.1680653	Qkm_2–6	−0.4187837	V_7	1.061520	1.061523
P_11	−0.1932750	Qkm_3–4	0.4352321	V_8	1.090000	1.090108
P_13	−0.2724101	Qkm_3–6	0.1954689	V_9	1.055932	1.055945
Q_1	0.2622563	Qkm_5–6	0.3446874	V_10	1.050985	1.050994
Q_2	0.1661884	Qkm_5–10	−0.0567783	V_11	1.056907	1.056914
Q_3	0.2858760	Qkm_6–7	−0.5718472	V_12	1.055189	1.055210
Q_5	0.0521873	Qkm_7–13	−0.3736071	V_13	1.050382	1.050400
Q_6	−0.4428740	Qkm_7–14	0.0413541	V_14	1.035530	1.035515
Q_8	0.5188807	Qkm_11–12	−0.4255870	Θ_1	0.000000	0.000000
Q_9	−0.2623013	Qkm_13–14	−0.0298895	Θ_2	−4.982589	−4.982760
Q_11	−0.3962582	Qkm_14–15	0.0853592	Θ_3	−12.72510	−12.72724
Q_13	−0.4587539	Qmk_6–2	−0.2531934	Θ_4	−10.31290	−10.31004
Pkm_2–3	−0.3717061	Qmk_4–3	0.0794752	Θ_5	−8.773854	−8.772070
Pkm_2–6	0.5311124	Qmk_5–3	0.5809284	Θ_6	−14.22095	−14.21995
Pkm_3–4	0.0146402	Qmk_10–5	0.0506414	Θ_7	−13.35963	−13.35709
Pkm_3–6	−0.4700330	Qmk_7–6	−0.0465348	Θ_8	−13.35963	−13.35709
Pkm_5–6	0.6365612	Qmk_12–7	0.3967287	Θ_9	−14.93852	−14.93650

Continued

Table 8.1 Estimated step no. 10: initial stationary condition—cont'd

Measurement	Normalized innovation	Measurement	Normalized innovation	State variable	Reference value	Estimated value
Pkm_5–10	0.7110650	Qmk_13–7	−0.7053349	Θ_10	−15.09729	−15.09693
Pkm_6–7	0.2945354	Qmk_9–8	−0.4627601	Θ_11	−14.79062	−14.78984
Pkm_7–13	−0.6031647	Qmk_10–8	0.0681805	Θ_12	−15.07558	−15.07365
Pkm_7–14	0.3370787	V_1	−0.6259802	Θ_13	−15.15628	−15.15520
Pkm_11–12	−0.2734427	V_3	−0.5870242	Θ_14	−16.03364	−16.03472
Pkm_13–14	−0.0367685	V_4	0.7317666			
Pkm_14–15	0.5376321	**V_5**	**−1.8943582**	Asymmetry index		
Pmk_6–2	−0.6431915	**V_8**	**1.2454309**	−0.3417407		
Pmk_4–3	−0.7288150	V_9	−0.8367944			
Pmk_5–3	0.0383584	V_11	−0.8333578			
Pmk_10–5	0.6123081	V_13	1.0668362			

Table 8.2 Estimated step no. 30: In the presence of the load ramp.

Measurement	Normalized innovation	Measurement	Normalized innovation	State variable	Reference	Estimated
P_1	1.6803673	Pmk_7–6	−1.7019879	V_1	1.060000	1.059739
P_2	−1.5241814	Pmk_12–7	−1.1194021	V_2	1.045000	1.044906
P_3	0.6336098	Pmk_13–7	−2.2851799	V_3	1.010000	1.010047
P_5	−0.0802524	Pmk_9–8	−0.0000056	V_4	1.017145	1.017260
P_6	−2.1016755	Pmk_10–8	0.7245799	V_5	1.019048	1.019142
P_8	−0.0000056	Qkm_2–3	0.7566512	V_6	1.070000	1.070009
P_9	−0.3715524	**Qkm_2–6**	**4.4732413**	V_7	1.060945	1.061000
P_11	−0.4523261	Qkm_3–4	1.4817686	V_8	1.090000	1.090010
P_13	−1.3767190	Qkm_3–6	1.8918506	V_9	1.055026	1.055071
Q_1	0.9441078	Qkm_5–6	1.6417012	V_10	1.050081	1.050145
Q_2	1.4767676	Qkm_5–10	0.1712470	V_11	1.056373	1.056423
Q_3	−0.7852853	Qkm_6–7	0.0365529	V_12	1.054835	1.054930
Q_5	0.4628488	Qkm_7–13	−0.2420315	V_13	1.049881	1.049962
Q_6	0.7208781	Qkm_7–14	−0.2471304	V_14	1.034414	1.034524
Q_8	−0.3402248	Qkm_11–12	−0.0764616	Θ_1	0.000000	0.000000
Q_9	−0.4084272	Qkm_13–14	−0.1112488	Θ_2	−5.047137	−5.012938
Q_11	0.0159025	Qkm_14–15	−0.1054920	Θ_3	−12.83494	−12.77753
Q_13	0.2764148	Qmk_6–2	2.8861973	Θ_4	−10.45251	−10.37670
Pkm_2–3	1.6029444	Qmk_4–3	1.2147318	Θ_5	−8.896483	−8.831422
Pkm_2–6	0.3396686	**Qmk_5–3**	**3.4588988**	Θ_6	−14.43929	−14.31961
Pkm_3–4	0.8924150	Qmk_10–5	1.7282845	Θ_7	−13.55923	−13.43143
Pkm_3–6	2.3397812	Qmk_7–6	1.1539151	Θ_8	−13.55923	−13.43143
Pkm_5–6	−1.1858575	Qmk_12–7	0.3291984	Θ_9	−15.16975	−15.04690

Continued

Table 8.2 Estimated step no. 30: In the presence of the load ramp—cont'd

Measurement	Normalized innovation	Measurement	Normalized innovation	State variable	Reference	Estimated
Pkm_5–10	2.2035064	Qmk_13–7	0.4911245	Θ_10	−15.33205	−15.20539
Pkm_6–7	1.8230484	Qmk_9–8	0.3991363	Θ_11	−15.01945	−14.89485
Pkm_7–13	1.2634323	Qmk_10–8	0.0942691	Θ_12	−15.31181	−15.17909
Pkm_7–14	1.6902032	V_1	−1.4223555	Θ_13	−15.39351	−15.26131
Pkm_11–12	−0.6722381	V_3	−2.0305786	Θ_14	−16.28939	−16.15255
Pkm_13–14	0.0341963	V_4	0.5173117			
Pkm_14–15	1.2184416	V_5	−0.3608499	Asymmetry index		
Pmk_6–2	−1.3859677	V_8	−0.0705017	0.4872715		
Pmk_4–3	−0.2641632	V_9	1.0196515			
Pmk_5–3	−0.4999656	V_11	0.5245368			
Pmk_10–5	−1.4240617	V_13	−0.2139842			

Table 8.3 Estimated step no. 40: In the presence of the load ramp and GEs (at Pkm_3–4 and Qkm_3–4).

Measurement	Normalized innovation	Measurement	Normalized innovation	State variable	Reference	Estimated
P_1	1.6424649	Pmk_7–6	−2.0263388	V_1	1.060000	1.059524
P_2	−3.8663275	Pmk_12–7	−0.5044781	V_2	1.045000	1.044622
P_3	5.9944160	Pmk_13–7	−2.0121470	V_3	1.010000	1.008464
P_5	−0.1844112	Pmk_9–8	−0.0000118	V_4	1.016606	1.016650
P_6	−1.9668509	Pmk_10–8	0.4986327	V_5	1.018570	1.018564
P_8	−0.0000118	Qkm_2–3	−0.3944672	V_6	1.070000	1.069844
P_9	0.0540592	Qkm_2–6	5.3929934	V_7	1.060356	1.060310
P_11	−0.6771246	**Qkm_3–4**	**7.4901245**	V_8	1.090000	1.089898
P_13	−0.8588791	Qkm_3–6	3.5133721	V_9	1.054097	1.054049
Q_1	0.7081545	Qkm_5–6	1.3976899	V_10	1.049155	1.049056
Q_2	−0.7663617	Qkm_5–10	0.3041864	V_11	1.055827	1.055715
Q_3	−1.1220622	Qkm_6–7	0.5144148	V_12	1.054473	1.054412
Q_5	0.7054467	Qkm_7–13	−1.4090234	V_13	1.049368	1.049315
Q_6	1.7563450	Qkm_7–14	−0.2948226	V_14	1.033271	1.033267
Q_8	0.1867523	Qkm_11–12	0.3588705	Θ_1	0.000000	0.000000
Q_9	−0.3950927	Qkm_13–14	−0.2169057	Θ_2	−5.113114	−5.089535
Q_11	0.4147946	Qkm_14–15	−0.0784058	Θ_3	−12.94721	−13.07766
Q_13	0.5487735	Qmk_6–2	3.7442416	Θ_4	−10.59516	−10.48686
Pkm_2–3	1.6323411	Qmk_4–3	3.9411416	Θ_5	−9.02179	−8.93380
Pkm_2–6	1.3379989	Qmk_5–3	5.4668019	Θ_6	−14.66250	−14.50128
Pkm_3–4	**22.4846595**	Qmk_10–5	3.3321796	Θ_7	−13.76325	−13.58554
Pkm_3–6	1.5766445	Qmk_7–6	1.8810096	Θ_8	−13.76325	−13.58554
Pkm_5–6	−2.4458518	Qmk_12–7	0.0679103	Θ_9	−15.40612	−15.23762

Continued

Table 8.3 Estimated step no. 40: In the presence of the load ramp and GEs (at Pkm_3–4 and Qkm_3–4)—cont'd

Measurement	Normalized innovation	Measurement	Normalized innovation	State variable	Reference	Estimated
Pkm_5–10	1.8597402	Qmk_13–7	0.2758347	Θ_10	−15.57203	−15.40470
Pkm_6–7	2.0962922	Qmk_9–8	−0.0754790	Θ_11	−15.25337	−15.08815
Pkm_7–13	2.6099130	Qmk_10–8	0.0814493	Θ_12	−15.55329	−15.37804
Pkm_7–14	1.2941521	V_1	0.4920113	Θ_13	−15.63602	−15.46032
Pkm_11–12	−0.8757630	V_3	−0.1021971	Θ_14	−16.55086	−16.36262
Pkm_13–14	0.0839511	V_4	−0.3992350			
Pkm_14–15	0.5864872	V_5	0.5128121	Asymmetry index		
Pmk_6–2	−1.9467423	V_8	0.1885211	4.1762113		
Pmk_4–3	3.3682234	V_9	−0.7986730			
Pmk_5–3	−1.9875676	V_11	−0.8425770			
Pmk_10–5	−1.5213570	V_13	0.9002423			

Table 8.4 Estimated step no. 70: IN the presence of the sudden load increase in four nodes.

Measurement	Normalized innovation	Measurement	Normalized innovation	State variable	Reference	Estimated
P_1	6.5135741	Pmk_7–6	−4.0840386	V_1	1.060000	1.058485
P_2	−5.9122247	Pmk_12–7	−0.9436213	V_2	1.045000	1.044409
P_3	−5.0711238	Pmk_13–7	−3.6770754	V_3	1.010000	1.010614
P_5	−0.1283272	Pmk_9–8	−0.0000541	V_4	1.014740	1.015041
P_6	−4.0271774	Pmk_10–8	0.4157290	V_5	1.016873	1.017039
P_8	−0.0000541	Qkm_2–3	−0.7961194	V_6	1.070000	1.069963
P_9	0.2431524	**Qkm_2–6**	**15.8701274**	V_7	1.058624	1.058831
P_11	−1.0798324	Qkm_3–4	14.1186307	V_8	1.090000	1.090093
P_13	−3.1278343	Qkm_3–6	6.6798790	V_9	1.051511	1.051688
Q_1	3.8378652	Qkm_5–6	6.1734422	V_10	1.046685	1.046698
Q_2	7.4948214	Qkm_5–10	0.0786221	V_11	1.054413	1.054413
Q_3	8.6290821	Qkm_6–7	0.0865731	V_12	1.053305	1.053546
Q_5	1.4037598	Qkm_7–13	−0.1496097	V_13	1.047317	1.047559
Q_6	2.2765046	Qkm_7–14	−0.1623035	V_14	1.029074	1.029501
Q_8	0.6143918	Qkm_11–12	0.3743756	Θ_1	0.000000	0.000000
Q_9	−0.1123756	Qkm_13–14	0.0038668	Θ_2	−5.484951	−5.254351
Q_11	0.3644774	Qkm_14–15	−0.1910796	Θ_3	−13.79604	−13.27172
Q_13	1.4041959	**Qmk_6–2**	**14.8922366**	Θ_4	−11.32016	−10.88658
Pkm_2–3	6.5302480	Qmk_4–3	14.6539146	Θ_5	−9.636424	−9.279717
Pkm_2–6	5.8461839	Qmk_5–3	14.0909809	Θ_6	−15.61566	−15.10848
Pkm_3–4	4.4987013	Qmk_10–5	4.9080855	Θ_7	−14.65621	−14.13169
Pkm_3–6	5.1727756	Qmk_7–6	5.1646549	Θ_8	−14.65621	−14.13169
Pkm_5–6	−6.6001754	Qmk_12–7	0.2524145	Θ_9	−16.38718	−15.87004

Continued

Table 8.4 Estimated step no. 70: IN the presence of the sudden load increase in four nodes—cont'd

Measurement	Normalized innovation	Measurement	Normalized innovation	State variable	Reference	Estimated
Pkm_5–10	2.4490911	Qmk_13–7	0.7009807	Θ_10	−16.56036	−16.04434
Pkm_6–7	3.3336939	Qmk_9–8	0.4046862	Θ_11	−16.22854	−15.71493
Pkm_7–13	4.0574550	Qmk_10–8	−0.1291239	Θ_12	−16.56824	−16.02598
Pkm_7–14	3.5804862	V_1	−1.5296719	Θ_13	−16.66795	−16.12543
Pkm_11–12	−1.2150921	V_3	−0.5688249	Θ_14	−17.65802	−17.08801
Pkm_13–14	0.3792664	V_4	−1.0203802			
Pkm_14–15	2.6227415	V_5	1.2954986	Asymmetry index		
Pmk_6–2	−5.1901487	V_8	−0.0674302	1.1496038		
Pmk_4–3	−5.3176416	V_9	−0.4769760			
Pmk_5–3	−4.3338606	V_11	1.3500313			
Pmk_10–5	−2.5741797	V_13	0.3090042			

Table 8.5 Estimated step no. 90: In the presence of the sudden load decrease in four nodes.

Measurement	Normalized innovation	Measurement	Normalized innovation	State variable	Reference	Estimated
P_1	−6.1600658	Pmk_7–6	2.0930172	V_1	1.060000	1.060975
P_2	5.8393474	Pmk_12–7	0.7020379	V_2	1.045000	1.045137
P_3	5.5107323	Pmk_13–7	2.8480608	V_3	1.010000	1.008771
P_5	0.0023361	Pmk_9–8	0.0000185	V_4	1.015484	1.014971
P_6	2.7150665	Pmk_10–8	0.7537934	V_5	1.017573	1.017180
P_8	0.0000185	Qkm_2–3	−0.3683011	V_6	1.070000	1.069691
P_9	−0.2518494	**Qkm_2–6**	**−13.9699213**	V_7	1.059132	1.058674
P_11	0.6528138	**Qkm_3–4**	**−17.2340681**	V_8	1.090000	1.089649
P_13	2.1926582	Qkm_3–6	−5.6439935	V_9	1.052170	1.051711
Q_1	−3.2751873	Qkm_5–6	−4.1920210	V_10	1.047234	1.046841
Q_2	−6.6545087	Qkm_5–10	−0.0168282	V_11	1.054695	1.054330
Q_3	−4.4454122	Qkm_6–7	0.2804444	V_12	1.053725	1.053252
Q_5	−1.0750497	Qkm_7–13	0.3082287	V_13	1.048307	1.047835
Q_6	−1.4323006	Qkm_7–14	0.4107481	V_14	1.030903	1.030282
Q_8	−0.3306288	Qkm_11–12	0.4087861	Θ_1	0.000000	0.000000
Q_9	1.1494830	Qkm_13–14	0.0852992	Θ_2	−5.249492	−5.431319
Q_11	0.0037336	Qkm_14–15	−0.2789969	Θ_3	−13.17929	−13.61818
Q_13	−0.4661318	Qmk_6–2	−11.4015038	Θ_4	−10.88992	−11.21420
Pkm_2–3	−6.1481934	Qmk_4–3	−10.7711264	Θ_5	−9.280719	−9.547965
Pkm_2–6	−3.9770020	Qmk_5–3	−9.8994352	Θ_6	−15.12402	−15.49342
Pkm_3–4	−4.3961336	Qmk_10–5	−2.0138340	Θ_7	−14.18491	−14.53725
Pkm_3–6	−2.9818962	Qmk_7–6	−2.4616513	Θ_8	−14.18491	−14.53725
Pkm_5–6	4.1067723	Qmk_12–7	−0.0898905	Θ_9	−15.89479	−16.26370

Continued

Table 8.5 Estimated step no. 90: In the presence of the sudden load decrease in four nodes—cont'd

Measurement	Normalized innovation	Measurement	Normalized innovation	State variable	Reference	Estimated
Pkm_5–10	−1.6762268	Qmk_13–7	−0.4098634	Θ_10	−16.06819	−16.44447
Pkm_6–7	−2.7355267	Qmk_9–8	−0.2674908	Θ_11	−15.73699	−16.11083
Pkm_7–13	−3.0114116	Qmk_10–8	−0.2784584	Θ_12	−16.05252	−16.44417
Pkm_7–14	−2.5356518	V_1	1.0442430	Θ_13	−16.13738	−16.52888
Pkm_11–12	0.6592683	V_3	−1.4276833	Θ_14	−17.09155	−17.50469
Pkm_13–14	−0.2972686	V_4	−0.7414914			
Pkm_14–15	−2.1040836	V_5	0.9204354	Asymmetry index		
Pmk_6–2	4.9231833	V_8	−0.8577698	−1.5022741		
Pmk_4–3	4.5718486	V_9	−0.2813155			
Pmk_5–3	4.1564038	V_11	−0.2703178			
Pmk_10–5	1.0389264	V_13	1.8268842			

Since the asymmetry index keeps its value below the threshold, we can discard the hypothesis of GE in the measurement. The conclusion is for a sudden change in the loading, in this case due to the presence of a load ramp.

The normalized innovation above the detection threshold value, for measurements *Pkm_3–4* and *Qkm_3–4,* indicates the presence of an anomaly at this step. Since the absolute value of the asymmetry index is also above its detection threshold, the conclusion is for the presence of GEs. The measurement with the largest normalized innovation is accurately identified in this case. It demonstrates the association of the normalized innovation and the asymmetry index to reinforce the bad data processing in the context of dynamic state estimation.

In this case, once again, the normalized innovation above the detection threshold, for several measurements, indicates the presence of an anomaly at this step. Since the asymmetry index keeps its value below the threshold, the hypothesis of GE in the measurement is discarded. The conclusion is for a sudden change in the loads, in this case a more abrupt event due to the presence of a sudden load change (in this case the load increase).

Like the results in the previous table, once again, the normalized innovation above the detection threshold, for several measurements, indicates the presence of an anomaly in this time step. Since the asymmetry index keeps its value below the threshold, the hypothesis of GE in the measurement is discarded. The conclusion is thus a sudden change in the loads, in this case also from a more abrupt event due to the presence of a sudden load change (in this case the load decrease).

8.7.2.2 Adaptive KF dynamic state estimator

As presented in the previous results, the presence of GEs sensibly reduces the accuracy of the estimation. Besides, sudden load changes also distort the results, since such a nonstationary transition is not contemplated within the model of the temporal relations between states (the assumption that the previous state is a good approximation for the current one). Nonetheless, in such situations, the association of the normalized innovation and the asymmetry index provide the necessary information to discriminate between such two different anomalies.

To surpass the harmful effects of these anomalies in the estimation process, a simple but effective adaptation strategy can be performed. The goal is to restore the accuracy of the KF and provide the best estimation possible for the state of the power system. The adaptive strategy consists of two additional steps for the estimation process:

1. If a GE is detected (if $|\lambda_{k,\ i}| > a$ and $\gamma_k > |\gamma_{max}|$), the measurement with the largest normalized innovation is discarded, and the estimation process is repeated without the discarded measurement.
2. If an anomaly is detected (if $|\lambda_{k,\ i}| > a$ and $\gamma_k < |\gamma_{max}|$), the estimation restarts its initial condition by neglecting the time relations between states, that is, using only the current snapshot of measurements, and the estimation process is repeated using a static estimator to provide the estimated state.

The procedure continues until any GE or anomaly is detected by the presented method. The estimated state is then provided as the final solution whenever the effects of both GEs and anomalies are suppressed. By employing this adaptive strategy, the KF can regain its accuracy under the influence of GEs and anomalies in the power grid.

In order to illustrate the effect of such an adaptive strategy, Fig. 8.9 presents the estimated state variables for node 14 in comparison to the reference load flow of the system. Fig. 8.10 presents the estimation error boxplot between the estimated state variables and the reference values obtained in the load flow calculation for all nodes. The effect of the adaptive strategy successfully reduces the negative effects of the anomalies in the estimation. In case of GE, by neglecting the contaminated measurement, the dynamic state estimator readily regains its accuracy.

During the presence of a load ramp and the sudden load change, the strategy comprises of a temporarily change toward a static estimation, neglecting the system dynamics in order to capture the transition that occurred on the network. As it can be seen, during the load ramp, the adaptation strategy is successively triggered more than once, at time steps 26 s, 35 s, 44 s, and 54 s. This is because the load ramp is a persistent anomaly over 40 s, that continuously changes the system dynamics during this interval. Meanwhile, the sudden load change (load increase and load decrease) is abrupt changes that occur in a single instant, at time step 70 s and 90 s. In this case, the anomaly needs to be suppressed only once, and the dynamic state estimator can slowly return to the correct and new stationary level after the adaptation. Overall, this adaptation strategy is enough to keep track of the state variable automatically, once the detection, identification, and suppression strategy are incorporated in the same framework, triggered by the normalized innovation and asymmetry index, increasing estimation accuracy.

The effect of the adaptation strategy is further demonstrated in Figs. 8.11 and 8.12, with the maximum normalized innovation and asymmetry index for each instant. The results are compared in both figures with the previously

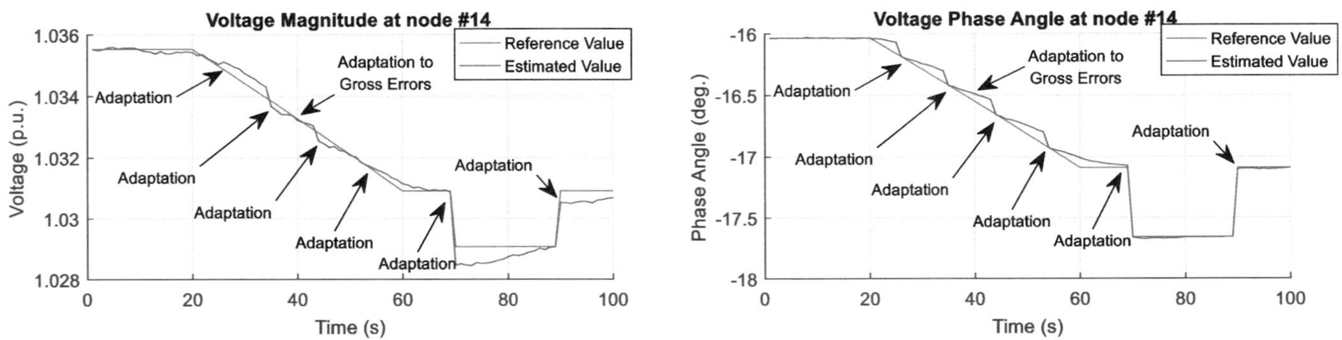

Fig. 8.9 Example of estimated state variables (voltage magnitude and phase angle) at bus 14 during the simulations with the adaptive dynamic state estimator.

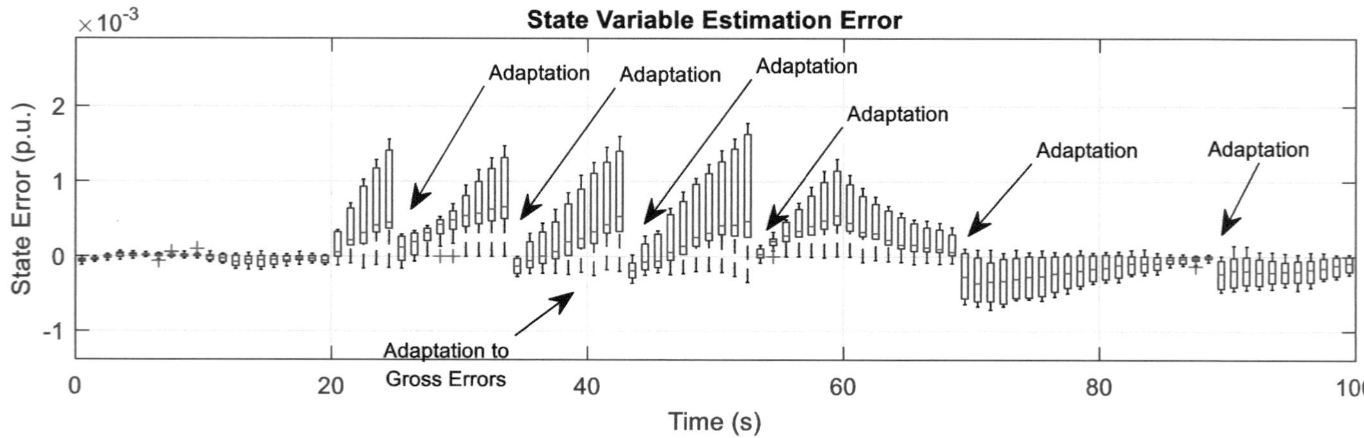

Fig. 8.10 Boxplot of estimation error in all state variables at each instant of the simulation, using the adaptive Kalman filter approach. The estimation error was obtained from the difference between the load flow reference values and the final estimated state variables.

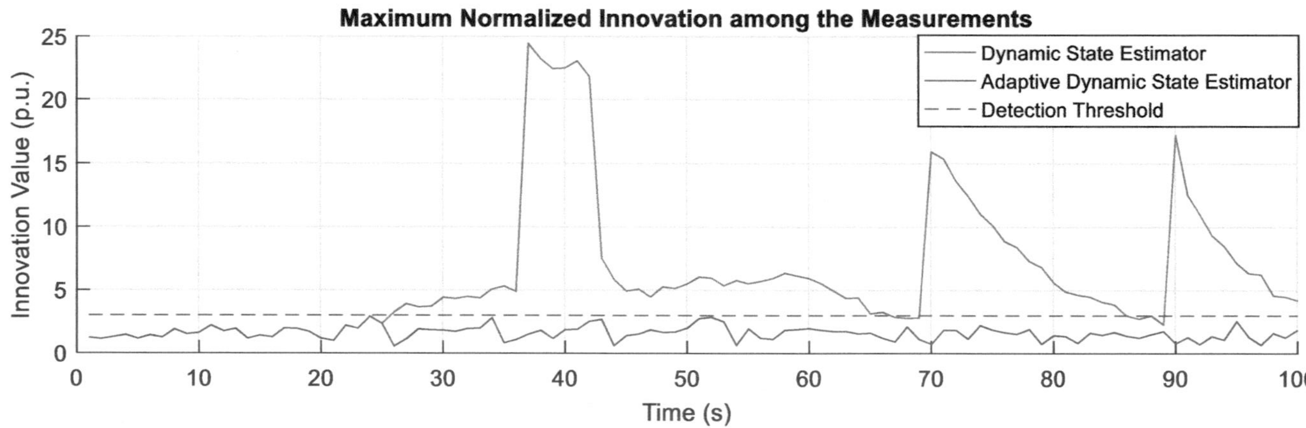

Fig. 8.11 Maximum normalized innovation index in the IEEE 14 bus test system during the simulations for the adaptive dynamic state estimator.

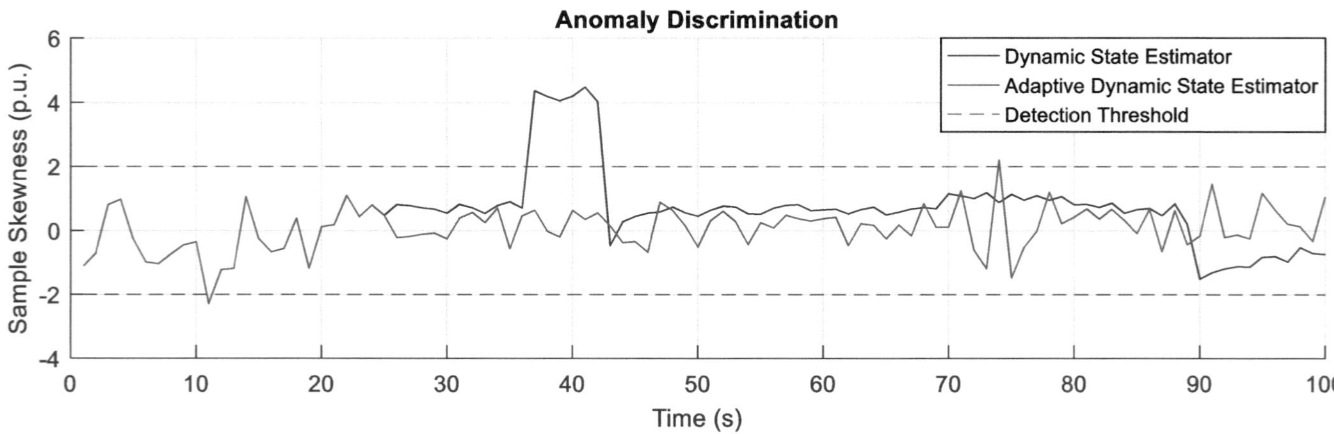

Fig. 8.12 Asymmetry index from the sample skewness in the IEEE 14 bus test system during the simulations for the adaptive dynamic state estimator.

obtained results, that is, without the adaptation strategy. After the adaptation steps are performed, the normalized innovation keeps its value below the detection threshold during the entire simulation. It shows that the dynamic state estimator, associated with the adaptive strategy, can fully detect, identify, and suppress the effects of anomalies in both measurement set and system dynamics.

8.7.3 Case 2—Dynamic state estimation in the presence of random load variation and different level of GEs

Case summary: Estimation corresponding to 100s of data, from different time instant, obtained in a sequence, with introduction of measurement GE.

The data to perform the GE analysis will be the same used for the case of anomaly detection case. The IEEE 30 bus transmission test system is used for the tests in this case.

The simulation was carried out, in a similar manner to the previous section, based on the reference load flow scenario of the IEEE 30 bus test system (Christie, 1993). In this case, a random load variation was included with a uniform variation of 1% in all loads. Besides, a sudden load increase of 5.0% occurs from $t=40s$ to $t=90s$ at six specific nodes (nodes 3, 4, 15, 19, 21, and 30). Fig. 8.13 illustrates the total net load and total generation of the IEEE 30 system during this sequence of events.

The measurements considered in the simulations are shown in Table 8.6.

The simulation consists on a similar setup as described in the previous section, where measurement values present random Gaussian error of 2% for active and reactive power measurements and 1% for voltage magnitude measurement. In this simulation, different values of GEs were inserted in different measurements according to the respective measurement standard deviation:

1. GE of five times the standard deviation (5σ) is introduced at measurement P_{2-4} and Q_{2-4} from $t=13$s to $t=14$s (during 2 s);
2. GE of 10 times the standard deviation (10σ) is introduced at measurement P_{10-21} and Q_{10-21} from $t=23$s to $t=24$s (during 2 s);
3. GE of 15 times the standard deviation (15σ) is introduced at measurement P_{25-27} and Q_{25-27} from $t=43$s to $t=44$s (during 2 s);
4. GE of 20 times the standard deviation (20σ) is introduced at measurement V_8 from $t=53$s to $t=54$s (during 2 s);
5. GE of 25 times the standard deviation (25σ) is introduced at measurement V_{27} from $t=73$s to $t=74$s (during 2 s).

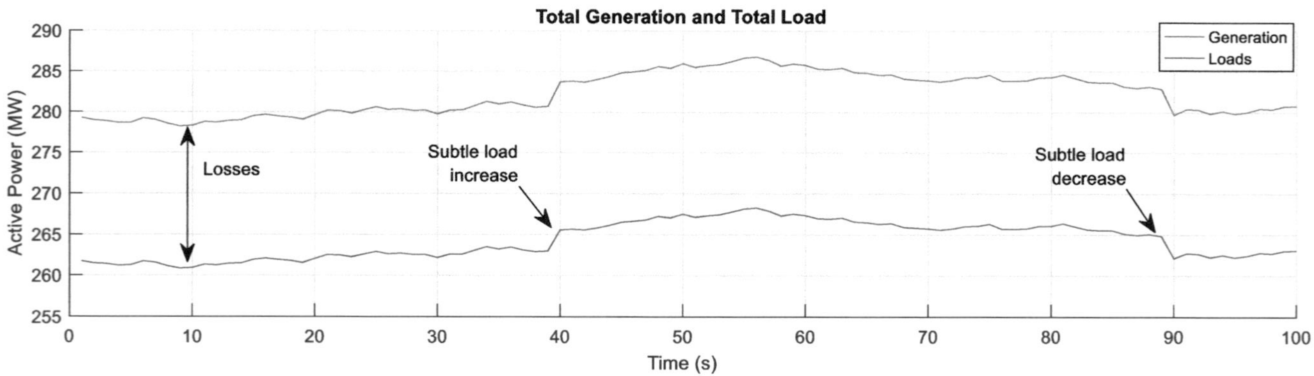

Fig. 8.13 Total net load and generation in the IEEE 30 bus test system for the time period of the simulation.

Table 8.6 Measurement location for the IEEE 30 bus transmission test system.

Measurement location		
Active and reactive power injection at nodes	**Active and reactive power flow at branches**	**Voltage magnitude at node**
1;2;5;6;8;10;11;13;15;17;19 21;22;24;25;27;30	4–12;6–9;6–10;28–27;1–2;1-3;2–4;2–5 2–6;6–7;6–8;6–28;10–17;10–20 10–21;10–22;12–15;15–18;15–23 16–17;25–26;25–27;27–29;27–30 12–4;9–6;10–6;27–28;13–12;14–12 15–12;23–15;20–19;22–21;24–22 29–27;30–29	1;2;5;8;10;11;13;24;27
Total		
19	39	9

Likewise, in the dynamic state estimator process, noise covariance matrix (Q) was obtained considering a 0.03% of possible random variation relative to the previous values of the state variables.

8.7.3.1 Analysis of results

For this case only, the detection indexes, the normalized innovation index along with the asymmetry index, are shown in the next figures and the summarized numerical results are presented in Table 8.7. Fig. 8.14 illustrates the maximum normalized innovation index from all measurements in each instant of the simulation and the detection threshold, what comprise the first step of the method to detect anomalies and GEs. Fig. 8.15 illustrates the calculated asymmetry index in each instant of the simulation and the detection threshold, what comprise the second step of the method to discriminate GEs.

Table 8.7 Summarized results for the IEEE 30 bus test system for GE detection, discrimination, and identification in different location.

Time instant	Description	Maximum normalized innovation	Measurement location	Asymmetry index
$t=5s$	With random load variation of 1%	1.486	V_{13}	0.1814
$t=13s$	With random load variation of 1% and GE of 5σ at #P_{2-4} and #Q_{2-4}	5.116	P_{2-4}	3.3592
$t=23s$	With random load variation of 1% and GE of 10σ at #P_{10-21} and #Q_{10-21}	8.883	P_{10-21}	5.4298
$t=40s$	During the sudden load increase of 5% in few nodes	4.955	Q_{10-6}	0.2059
$t=43s$	After the load increase, random variation of 1% and GE of 15σ at #P_{25-27} and #Q_{25-27}	11.704	P_{25-27}	3.4918

Table 8.7 Summarized results for the IEEE 30 bus test system for GE detection, discrimination, and identification in different location—cont'd

Time instant	Description	Maximum normalized innovation	Measurement location	Asymmetry index
$t=50$s	After the load increase and random variation of 1%	5.188	P_{2-4}	1.5064
$t=53$s	After the load increase, random variation of 1% and GE of 20σ at #V_8	18.409	V_8	8.1034
$t=70$s	After the load increase and random variation of 1%	2.0645	V_{24}	0.4539
$t=73$s	After the load increase, random variation of 1% and GE of 25σ at #V_{27}	20.742	V_{27}	10.1455
$t=90$s	During the sudden load decrease of 5% in few nodes	5.212	Q_{10-6}	−0.2454
$t=95$s	After the events with random load variation of 1%	3.0778	P_{2-4}	−0.9878

As it can be seen, just after the inclusion of GEs in the measurements, both the normalized innovation and the asymmetry index increase above the threshold value, in all levels and locations of simulated GEs. Consequently, the conclusion is for the occurrence of GE. Looking for the normalized innovation, the measurement with error was properly identified as well, even in the case of the smallest level of GE, of just 5σ. Besides, despite the random load variation of 1% during the other instants, the anomaly detection was only triggered during the sudden load increase and load decrease. The asymmetry index properly discriminated the occurrence of these events, keeping low values of asymmetry, below the threshold. The results obtained from the adaptive state estimator are also shown in both

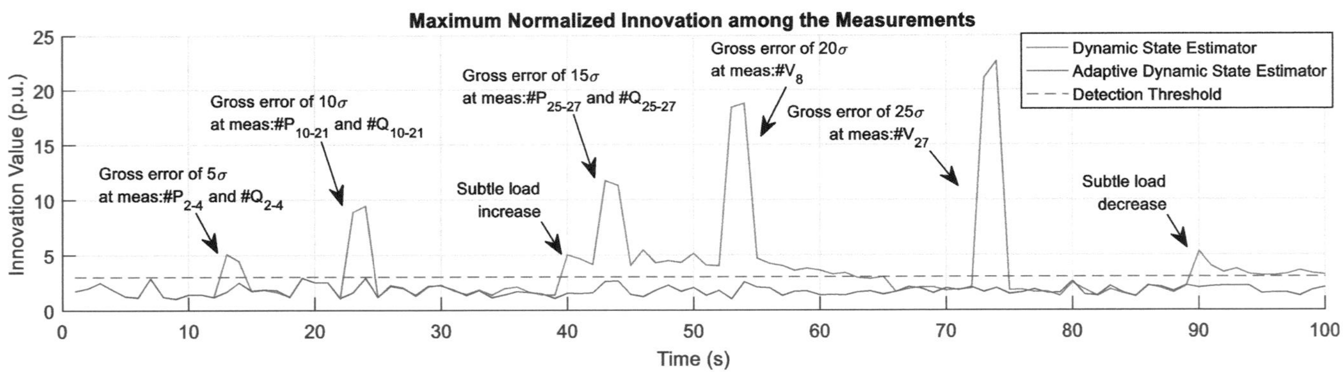

Fig. 8.14 Maximum normalized innovation index in the IEEE 30 bus test system during the simulations.

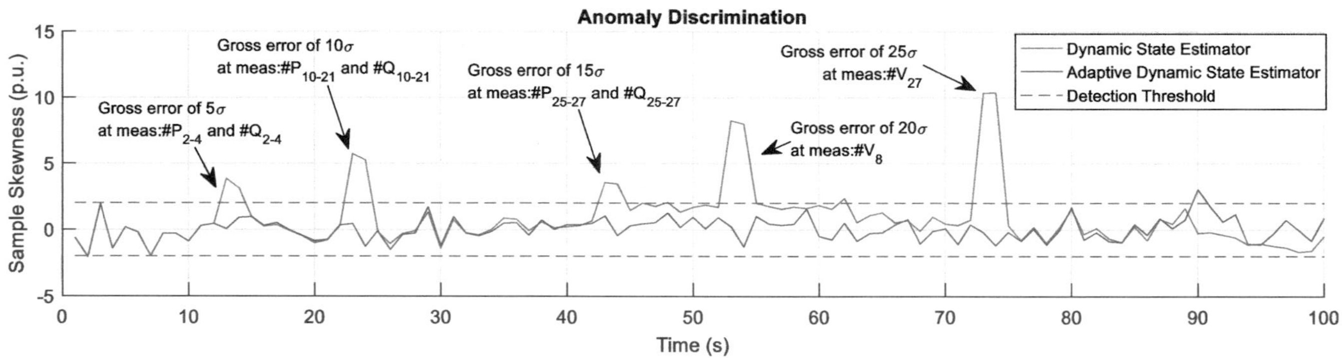

Fig. 8.15 Asymmetry index from the sample skewness in the IEEE 30 bus test system during the simulations.

figures, demonstrating the successful suppression of all cases of GEs and both sudden load changes.

Finally, the table below presents summarized results for a few instants of the simulation, in order to illustrate numerically the detection, identification, and discrimination processing of GEs. The largest normalized innovation and the respective measurement are shown as the first indicators of an anomaly during the estimation. The discrimination between GE and systemic anomalies is done by the calculated asymmetry index, also presented in the table. The conclusion about GEs in each instant is straightforward, as it can be seen in time instants 13 s, 23 s, 43 s, 53 s, and 73 s. By assessing both normalized innovation and asymmetry index, it was possible to detect and properly identify all levels of GEs in this simulation. Once again, two of the sudden load changes (load increase and load decrease) were also properly detected in the time instants 70 s and 90 s (Table 8.7).

8.7.4 Case 3—Dynamic state estimation in a medium-size power system with generation loss and multiple GEs

Case summary: Estimation corresponding to 100 s of data, from different time instant, obtained in a sequence, with introduction of measurement GE.

The data to perform the GE analysis will be the same used for the case of anomaly detection case.

For this final case study, the IEEE 118 transmission test system, a medium-size power system, was considered in a similar manner as the previous sections, based on the reference load flow scenario of the test system (Christie, 1993). The simulation comprises 100 sequential seconds, with the inclusion of 0.1% of random load variation in all nodes. This test scenario also considers a loss of generation at $t=40$s, simulated by the following event:

- Loss of generation at node 100, reducing 252 MW (6.1% of net load).

The measurement set is shown in Table 8.8.

The simulation consists on a similar setup as described in the previous section, where measurement values present random Gaussian error of 2% for active and reactive power measurements and 1% for voltage magnitude measurement. In this simulation, multiple GEs of 15 times the standard deviation (15σ) were inserted in multiple measurements in the same instant according to the following:

1. GE of 15 times the standard deviation (15σ) were introduced at measurement P_{30-17} and Q_{30-17}, P_{11-13} and Q_{11-13}, P_{56-57} and Q_{56-57}, P_{89-90} and

Table 8.8 Measurement location for the IEEE 118 bus transmission test system.

Measurement location		
Active and reactive power injection at nodes	**Active and reactive power flow at branches**	**Voltage magnitude at node**
6;17;18;20;22;23 25;26;28;30;31;45;46;47;48;53;59; 65;66;68;69;70;72;76;77;85;88;89;90; 91;95;100;103;105	1–2;6–7;8–9;8–5;4–11;12–14;12–16;19–20;15–33 34–37;40–41;34–43;46–48;49–50;49–51; 56–57	1;4;5;6;8;10;12;15;18;19;24;25 26;27;30;31;32
107;110;113	56–58;61–62;62–67;65–68;70–71;69–75;77–78 77–82;85–86;85–88;89–90;89–92; 92–93;80–96 80–97;92–102;42–4012-117;76–118;6–5;10–9 12–11;12–2;12–3;69–49; 15–14;18–17;27–115 19–18;19–15;25–26;31–17;32–27;34–19;36–35 36–34;40–37;40–39; 42–41;46–4549-45;54–53 56–55;59–54;61–59;61–60;59–63;61–64;66–49 66–65;69–47; 70–24;72–24;72–71;73–71;74–70 77–76;80–77;80–79;80–81;85–8387-86;89–85 89–88;91–90;85–84;92–91;99–80;100–92;100–98 100–99;100–94;105–104;110–109;105–103;111–110	34;36;37;38;40 42;44;45;46;48 49;54;55;56;59 61;62;64;65;66 68
Total		
37	108	67

Q_{89-90}, V_{48} and V_{74} from $t=25$s to $t=26$s during the initial stationary condition with 0.1% of random load variation.

2. GE of 15 times the standard deviation (15σ) were introduced at measurement P_{30-17} and Q_{30-17}, P_{11-13} and Q_{11-13}, P_{56-57} and Q_{56-57}, P_{89-90} and Q_{89-90}, V_{48} and V_{74} from $t=55$s to $t=56$s during after the event of loss of generation.

For this test scenario, in the dynamic state estimator process, noise covariance matrix (Q) was obtained considering a 1.0% of possible random variation relative to the previous values of the state variables. This value was chosen without any concern for proper tuning.

8.7.4.1 Analysis of results

Fig. 8.16 illustrates the maximum normalized innovation index from all measurements in each instant of the simulation and the detection threshold, what comprise the first step of the method to detect anomalies and GEs. Fig. 8.17 illustrates the calculated asymmetry index in each instant of the simulation and the detection threshold, what comprise the second step of the method to discriminate GEs.

Once again, just after the inclusion of multiple GEs in the measurements, both the normalized innovation and the asymmetry index increase above the threshold value. Therefore the conclusion is for the occurrence of GE. Looking at the normalized innovation, the measurement with error was properly identified as well. Besides, the anomaly detection was only triggered during the loss on generation. The asymmetry index properly discriminated the occurrence of the event of loss of generation from a GE, keeping low values of asymmetry, below the threshold. The results obtained from the adaptive state estimator are also shown in both figures, once again demonstrating the successful suppression of all cases of GEs and the sudden loss of generation.

In order to further explore and illustrate the asymmetry index, captured by the sample skewness of the normalized innovation, Fig. 8.18 shows the cumulative probability plot for this test during different instants. The plot is compared to a standard normal distribution in terms of cumulative probability. It clearly shows the asymmetry present in the normalized innovation sample in the presence of GEs. Despite the results during the initial stationary event, as well as during the loss of generation, show a heavy-tailed distribution, it clearly has more symmetry between the left and right sides of the curve in red, that is, when compared to the standard normal curve.

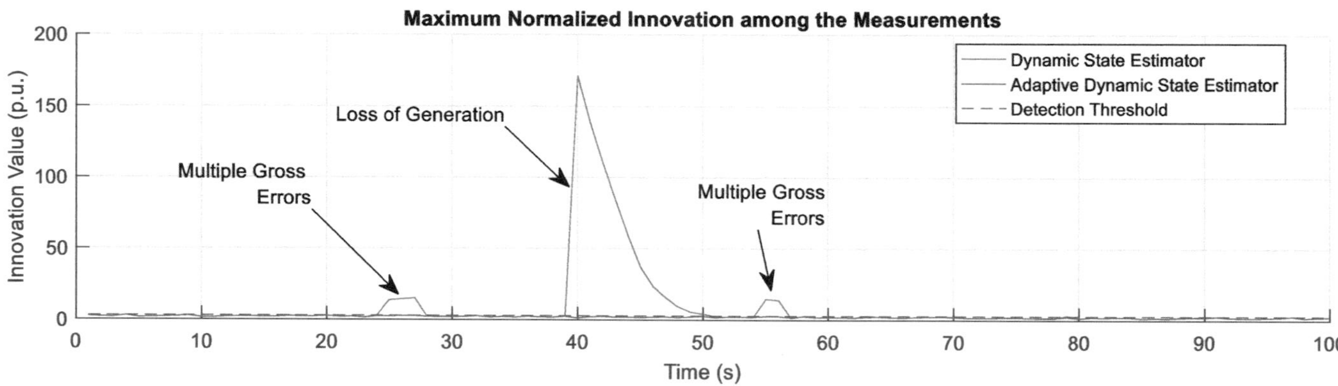

Fig. 8.16 Maximum normalized innovation index in the IEEE 118 bus test system during the simulations.

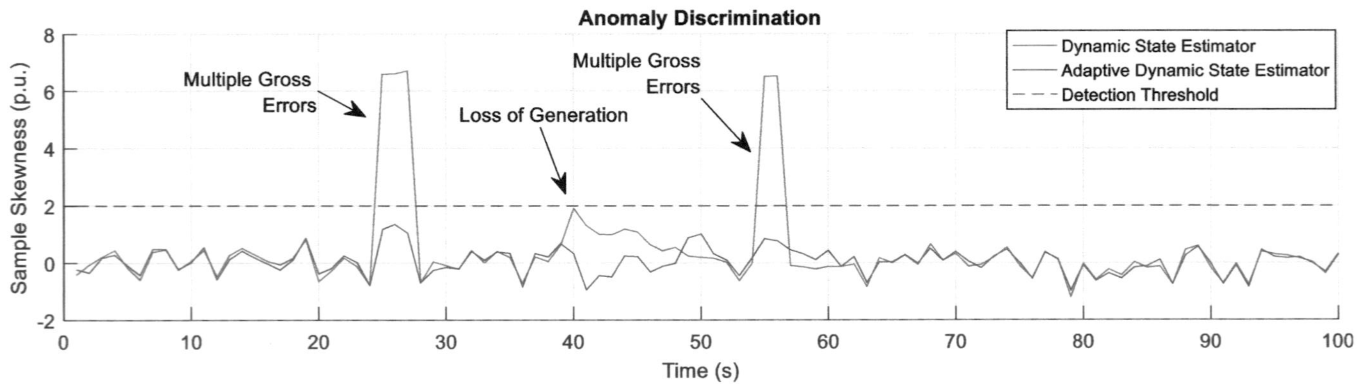

Fig. 8.17 Asymmetry index from the sample skewness in the IEEE 118 bus test system during the simulations.

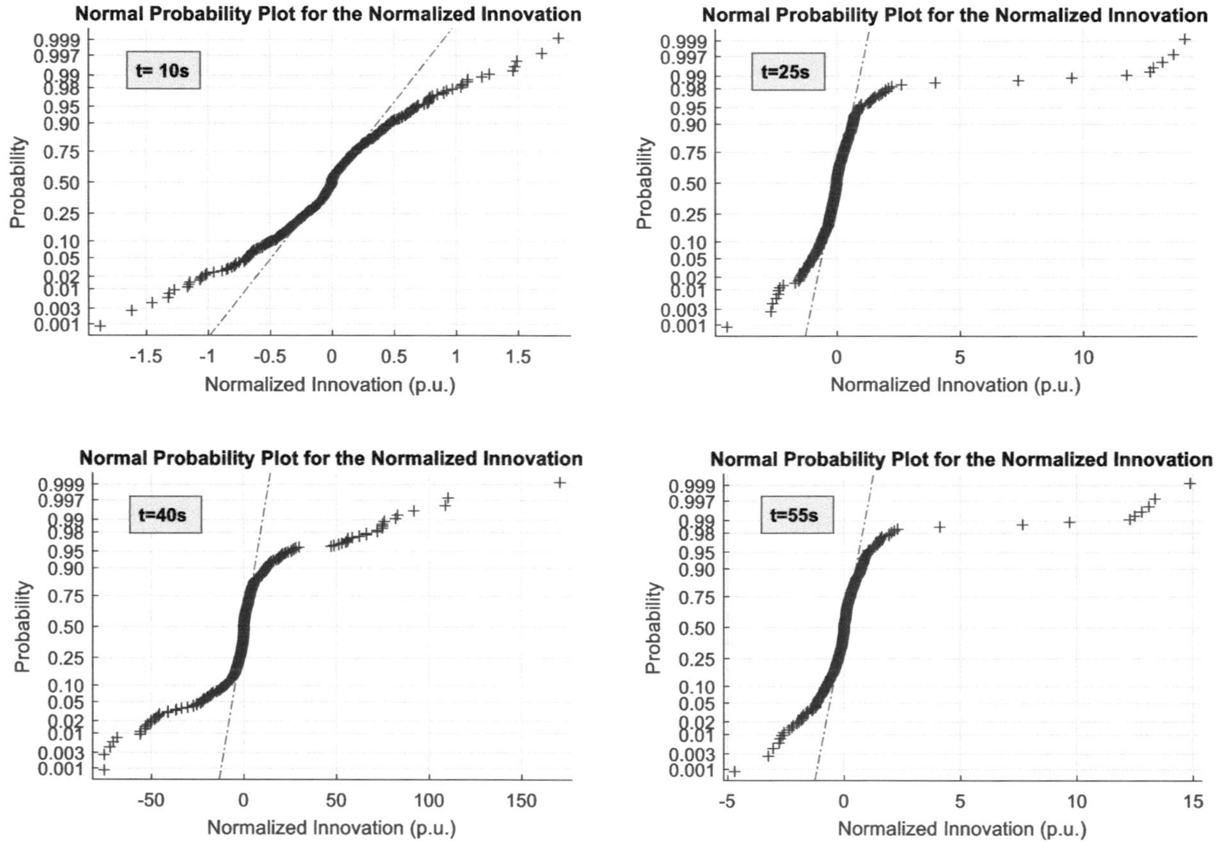

Fig. 8.18 Probability plots for the normalized innovation in different instants: (1) $t = 10$s in the initial stationary condition; (2) $t = 25$s with multiple GEs; (3) during the loss of generation event; and (4) after losing the generator and in the presence of multiple GEs.

Table 8.9 Summarized results for the IEEE 118 bus test system for GE detection, discrimination, and identification in different location.

Time instant	Description	Maximum normalized innovation	Measurement location	Asymmetry index
$t=10$s	Initial stationary condition with 0.1% of load random variation	1.858	V_{85}	0.0564
$t=25$s	With 0.1% of load random variation and in the presence of multiple GEs of 15σ	13.825	V_{74}	6.3732
$t=40$s	During the event of loss of generation at node 100	171.424	P_{77-82}	1.9158
$t=55$s	After the loss of generation, with 0.1% of load random variation and in the presence of multiple GEs of 15σ	14.987	V_{74}	6.2820
$t=95$s	Final steady state condition with 0.1% of load random variation	2.368	V_{37}	0.2947

Finally, Table 8.9 presents summarized results for a few instants of the simulation to illustrate numerically the detection, identification, and discrimination processing of GEs. The largest normalized innovation, the respective measurement, and the asymmetry index are presented in the table.

References

Bretas, N.G., 1986. Estimação de Estado: Alguns Modelos; Detecção e Identificação de Medidas com Erros Grosseiros (In Portuguese), Livre-Docente Thesis. University of São Paulo.

Bretas, N.G., 1989. An iterative dynamic state estimation and bad data processing. Int. J. Electr. Power Energy Syst. 11 (1), 70–74.

Bretas, N.G., Bretas, A.S., 2018. The extension of the Gauss approach for the solution of an overdetermined set of algebraic non-linear equations. IEEE Trans. Circ. Syst. Exp. Briefs 65, 1269–1273.

Christie, R., 1993. University of Washington Power Systems Test Case Archive. available at: https://labs.ece.uw.edu/pstca/.

Coutto Filho, M., Souza, J., 2009. Forecasting-aided state estimation part I: panorama. IEEE Trans. Power Syst. 24 (4), 1667–1677.

Gauss, C.F., 1795. Theoria Motus Corporum Coelestium in Sectionibus Conicis Solem Ambientum, Hamburg, 1809 (Translated as "Theory of the Motion of the Heavenly Bodies Moving about the Sun in Conic Sections"). Dover Publications, New York.

Kalman, R.E., 1960. A new approach to linear filtering and prediction problems. ASME J. Basic Eng. 82 (1), 35–45.

Kalman, R.E., Bucy, R.S., 1961. New results in linear filtering and prediction problem. ASME J. Basic Eng. 83 (1), 95–108.

Kolmogorov, A.N., 1940. Stationary Sequences in Hilbert Spaces, (in Russian), Moscow University Mathematics Bulletin, vol. 2 no. 61941. (translated by John Crerar Library, Chicago, 1978).

Krein, M.G., 1945. A generalization of the results of G. Szegö, V. I. Smirnov, and A. N. Kolmogorov. Doklady Akademii Nauk SSSR 46 (3), 95–98.

Legendre, A.M., 1805. Nouvelles Méthodes pour la Détermination des Orbites des Comètes. Firmin Didot, Paris.

Neugebauer, O., 1952. The Exact Sciences in the Antiquity. Princeton University Press.

Shivakumar, N., Jain, A., 2008. A Review of Power System Dynamic State Estimation Techniques. POWERCON - IEEE Power India Conference, New Delhi, India, pp. 1–6.

Wiener, N., 1964. Extrapolation, Interpolation, and Smoothing of Stationary Time Series With Engineering Applications. The MIT Press, Cambridge.

Wilson, G.T., 1972. The factorization of matricial spectral densities. SIAM J. Appl. Math. 23, 420–426.

Zhao, J., et al., 2019. Power system dynamic state estimation: motivations, definitions, methodologies, and future work. IEEE Trans. Power Syst. 34 (4), 3188–3198.

CHAPTER 9

Data-driven state estimation in electric power systems

This chapter aims to provide an introduction to data-driven model-based state estimators for real-time monitoring of the power grid, highlighting the structure and motivation that led the authors to accomplish it by organizing the following sections:

- *Section 1.1 Data-Driven models* describes the value of the data-driven state estimation solutions considering temporal and spatial characteristics for real-time monitoring of electric power systems (EPSs). The concepts and general framework of this process are introduced, contextualizing the development of data-driven state estimators, and the need for evolution to keep in line with new trends and current technologies. Data-driven linear regression models are introduced.
- *Section 1.2 The Ensemble CorrDet with Adaptive Statistics (ECD-AS) algorithm* is presented. The Ensemble CorrDet (ECD) algorithm is initially introduced. The ECD algorithm described uses the fixed estimated mean and covariance of normal samples after training, rather than adapting to the changing state of power systems. This means that when applied to a realistic load profile for power systems, ECD statistics will quickly become outdated and the performance of the algorithm will decrease. Thus, there is a need for new algorithms that can adapt to the real-time environment. An enhanced, adaptive, and responsive data-driven method for gross error analytics, called Ensemble CorrDet with Adaptive Statistics (ECD-AS), that accounts for the rapidly changing power system state is introduced.

9.1 Data-driven models

The objective of the weighted least squares state estimator (WLS-SE) is to find the best estimate for the vector of states $\hat{x}$ that minimizes the cost function $J(x)$. On the Gaussian approach, this cost function is the weighted sum of the squares of the measurement residuals. Geometrically, $J(x)$ is the

Cyber-Physical Power Systems State Estimation
https://doi.org/10.1016/B978-0-323-90033-1.00003-2

Euclidean norm in the vector space of the measurements. However, as proved on previous work (Bretas et al., 2013a), the residuals are not random independent variables, and do not represent the masked component of the error. To overcome these difficulties, the concept of Innovation was introduced for power system state estimation (PSSE) aiming to include an estimation of the masked component of the error.

The innovation of a measurement, in the context of static state estimation, is defined as the new information that this measurement contains, related to all other measurements, that is, the innovation is contained in the portion of the measurement that is independent of the other measurements. A measurement with a low *innovation index (II)* indicates that a large component of its error is not reflected in its residual. Consequently, even when those measurements have gross errors, their residuals will be small. The *II* does not depend directly on the measurement's magnitude, but on the characteristic of the linear solution. In addition, a measurement with a low *II* indicates that most of its error is "masked". Consequently, the measurement's error will present little contribution to the value of the residual, leading the classical residual-based methodologies to failure. To solve this difficulty, a new objective function to minimize the norm of weighted measurement errors instead of residuals was presented (Bretas and Bretas, 2018). Thus, the extension of the Gauss approach minimizes the weighted norm of the error.

For such, one needs to estimate the weighted norm of the error, and then to minimize it through the Newton-Raphson method. It has been shown that the measurement error, in the Gauss approach, is in the measurement direction and that the error has a unique decomposition: (i) the first component, which is orthogonal to the Jacobian range space, the residual; and (ii) the other, which is on the Jacobian range space. The latter is hidden in the Jacobian space when one minimizes the residual. The solution of $z = h(x) + e$, where $z \in \mathbb{R}^m$ is the measurement vector; $x \in \mathbb{R}^N$ is the state variable vector; $h: \mathbb{R}^N \rightarrow \mathbb{R}^m$, $(m > N)$ is a continuously nonlinear differentiable function; $e \in \mathbb{R}^m$ is the measurement error vector assumed having zero mean, standard deviation σ, and Gaussian probability distribution; and $N = 2n - 1$ is the number of unknown state variables to be estimated (n is the number of buses of the power system), on the Gauss approach, is the vector $\hat{x}$ obtained iteratively from the solution of the linearized equation:

$$z = Hx + e \tag{9.1}$$

where $H = \partial h / \partial x$ is the Jacobian of $h(x)$. The component of the error that belongs to the space spanned by H must be zero; otherwise, z is not optimal.

If the measurement error e is contained inside an acceptable ball of radius r, that is, less than a threshold predefined value, the classical least mean squares still may offer a reasonable solution. However, for real-life situations, measurements may have large errors. Then, a generalization of the weighted least squares approximation is required. The basic idea is to couple the two components of the error in the subspace of H and its orthogonal complement (denoted $e_U^{(i)}$ and $e_D^{(i)}$, respectively) by defining the II for the i-th measurement as being:

$$II_i = \frac{\left\| e_D^{(i)} \right\|_{R^{-1}}}{\left\| e_U^{(i)} \right\|_{R^{-1}}} \tag{9.2}$$

Eq. (9.2) is easily derived just using the definition of the norm, where R is a positive definite symmetric matrix. Consequently, knowing the projection matrix, the measurement II can be calculated; knowing this index, and the measurement residual, the $e_U^{(i)}$ error component can also be estimated and the measurement error e composed.

The limitation of the previous quasi-static measurement model is that it lags temporal information. While spatial characteristics are modeled, the quasi-static measurement model does not represent the dynamics of the system. One potential solution is to develop dynamic state estimation measurement models, as presented in Chapter 8. An obvious challenge is that one can model so much, and that models and model parameters might also be time-dependent. The use of hybrid data-driven physics-based approaches for modeling the grid, and thus state estimation, is a natural solution.

Challenges arise due to the size of the grid as well as to its dynamic characteristic. Decentralized solutions are thus suggested. However, how does one use knowledge from one area toward the other area? Several solutions, including particle swarm optimization, have been studied for this. Further, how does one use knowledge represented in models toward data-driven hybrid solutions, which are able to represent the system dynamics? Reinforcement learning has been presented toward this.

To increase system measurement redundancy, the use of synthetic measurements has been investigated, that is, measurements artificially created in the state estimation process, which do not depend of historical data, as pseudomeasurements, for overall bad data detection (BDD) enhancement (Bretas et al., 2019). However, there is an implicit trade-off that must be considered. If the degrees of freedom of the measurement model increase, the threshold value considering the same level of confidence also increases. Thus, synthetic measurements must add information, and not be redundant.

The obvious question then is how to create nonredundant synthetic measurements? If the synthetic measurements are to be created from the classical state estimation process, they will be redundant information when used in bad data analytics. Therefore, strategies to modify the state estimation process need investigation.

Initially, one can investigate the use of the measurement weight matrix for such, decoupling the creation of synthetic measurements for bad data analytics and state estimation. Classical state estimation suggests that the weight of a measurement in the WLS SE should be related to the variance of the measurement residual, which is related to the meter precision. However, if a measurement is in error, does the meter precision really matter? Of course the answer is no.

The natural answer is to consider that the measurements with higher magnitude value will attract the solution of the state estimation. Therefore, if this measurement is in error, the solution will be attracted to it. One potential approach is the use of the measurement covariance equal to a percentage of the measurement value during the bad data process. The idea is that during the bad data process, all measurements are to be considered in error, thus there is no meaning in considering the meter precision. After bad data processing, when no gross errors are detected, then the state estimation should consider the meter precision.

Still, the number and location of synthetic measurements needs definition. Because the number of synthetic measurements is inversely proportional to the bad data detection sensitivity and allocation is to be in areas where the test would fail, the optimal location and number of measurements need to be defined. The value of information relates the uncertainty associated with the relationships and their expected utility for task completion. This facilitates an optimal exploration of the action choices.

Thus, the value of synthetic measurements for allocation criteria is to be considered. For value, one might consider the use of the *II* for the synthetic measurement value metric. The *II* measures the amount of masked error a measurement has, thus the uncertainty associated. System areas might have different measurement scenarios, thus the BDD test may perform differently. For such, the use of the vulnerability index (*VI*) developed in (Ruben et al., 2020a) as well as the metric for synthetic measurement allocation can be considered.

A measurement *II* quantifies the amount of new information it adds to the measurement set. Thus, the inverse of *II* specifies the amount of undetectable component one measurement has. The *VI* otherwise quantifies vulnerability at the bus level rather than the measurement level. One would like to answer the following question: where should one allocate synthetic measurements?

The natural investigation would be in areas of low measurement redundancy. The question then is how to automatically identify in the state estimation process these areas, which are dynamic because system topology might change, meters are added and taken out for maintenance, etc. The identification of low measurement redundancy system areas considers first the most correlated measurements to critical and low *II* measurements as well as high *VI* buses because these are locations which, if in error, the detection test most likely will fail.

The *II* and *VI* metrics value the impact of bad data on measurements in the state estimation process. However, the error might be in the model parameter, or the model itself. The allocation of synthetic measurements should consider the impact of errors on all bad data processing, not only on measurements. Gross error can be further on static and/or dynamic data.

A metric for such impact needs definition. Initially, one can consider the use of the sensitivity matrix (Trevizan et al., 2018) to value the effect of parameter and measurement errors to the bad data analytics. The sensitivity matrix (Rossoni et al., 2018) shows the effect of a parameter error on BDD. The key is to develop models that relate the error and the parameters. This is not very difficult to derive; however, the sensitivity matrix is dependent on the current operation condition, which is previously unknown.

Relaxation solutions for this metric creation (Bretas et al., 2019) have been explored. On such, initially one supposes that the system topology and parameter are known and without error. This will allow the initial estimation of the operation condition as well as the estimation of the sensitivity matrix and new error metric. This further would allow a new allocation of synthetic measurements to be estimated and used for bad data analysis. Of course, the physics-based system model might be simultaneously in error in addition to the measurements.

To consider this hypothesis, an iterative solution has been presented (Zou et al., 2019). On such, one calculates the error metrics first, considering the relaxed assumptions, and then creates the synthetic measurements and processes the bad data analytics updating of the system model. Then, one recalculates the error metrics and processes the bad data analytics. This iterative process would repeat until no error is detected. Initial results on small systems are most encouraging (Zou et al., 2019).

One would like to answer the following question: what is the number of synthetic measurements that one should create artificially to improve the overall bad data analytics? Further, this estimation should be done with every new measurement scan because the physics-based system model is dynamic. It seems clear that only measurements with high value should be created, and that the allocation should be in areas where the BDD would fail. However, the number of synthetic measurements is also related to the sensitivity of the detection test. Thus, we have contradictory goals.

Solutions for such have been presented, and the natural one is a goal programming (GP) framework. GP is a multiobjective programming technique in which one tries to minimize the distance of the solution to previously defined goals (Ruben et al., 2020a). The GP framework is designed to optimize objectives that are contradictory in nature. The most common example of this is using a cost-based goal against a disutility-based one. Added costs will minimize disutility and vice versa. A GP framework can mathematically model the interdependence between problem objectives (Ferreira et al., 2015). It will be impossible to achieve all the objectives in the problem, so optimization is extremely valuable in this scenario.

One question one would like to answer is how can we leverage the data-driven reinforcement learning (DRL) diagnosis to improve the physics-based model analytic solution for BDD? It is clear that while the quasistatic physics-based model is unable to represent the temporal characteristics of the environment, DRL can. Otherwise the physics-based model does represent the spatio characteristics of the local environment. Thus we have complementary information, which one can explore.

An example of such is the augmented BDD, which uses synthetic measurements while also considering meter allocation and DRL information. The idea is to augment the measurement model with DRL data, then apply a Chi-square hypothesis test. This idea is plausible because independent of the source of the estimate, the assumptions of the measurement covariance are the same, thus the test can be applied. Several works have further theorized that this new information is not redundant and could provide additional information to BDD, providing a temporal flavor for the quasistatic model. Initial results (Trevizan et al., 2019b) are most encouraging in small-scale systems.

The Chi-square hypothesis test would be still dependent of the significance level chosen and on the assumptions made. One question one would like to answer is how to use the complementary characteristics of DRL and the physics-based model while not constraining the analyses to the analytic test constraints? One natural investigation presented is information fusion. The idea is to explore both types of information in a combined format without the implicit constraints of the physics-based model analytic test. The Choquet Integral (Choquet, 1954) to combine the data-driven and physics-based model anomaly detection information has been presented for such.

The Choquet Integral (*CI*) takes into consideration the confidence of the various anomaly detection methods to provide the best detection solution. This Integral is an aggregation operator based on fuzzy measures. Depending on the values of each element in the fuzzy measure, the *CI* can represent a variety of relationships and combinations among the information sources.

To address the problem of gross errors (GEs) in PSSE, much research has been done using the physics-model based framework, that is, using the known topology of the grid, along with real-time measurements, to estimate the voltages at each bus throughout the system (state variables). The results of PSSE are the state estimation itself, and the residual value for each real-time measurement used to complete the PSSE process. The basic approach for detecting GE is to analyze residual values using the common Chi-square test to determine if the residuals are statistically large for that set of measurement values (Monticelli, 1999). From there, the authors of (Handschin et al., 1975) explored the use of the largest normalized residual test as a detection method. Later on, the authors of (Abur and Expósito, 2004) expanded upon the Chi-square test, using the normalized residual within the test itself, and showed significant improvement.

Another significant improvement upon the Chi-square test came along with the geometrical interpretation of the measurement residuals, which was explored in (Bretas et al., 2013a). With the geometrical view of the PSSE process, it was proven that errors and residuals were in different mathematical spaces, and are different quantities. Through the geometrical interpretation of the measurement residuals, one can see that there exists an undetectable component of the measurement error that is not captured by the residual values. The *II* was developed to mathematically estimate this undetectable component, which was used then to compute the composed measurement error. The innovation concept was explored and expanded upon in a variety of papers (Bretas et al., 2011; Bretas et al., 2017a; Bretas and Bretas, 2018; Bretas et al., 2019; Bretas et al., 2013b).

Other potential methods of GE analytics, as false data injection (FDI) treatment, such as (Kosut et al., 2010; Bobba et al., 2010), deviate from the standard processes of utility companies without enough improvement upon Chi-square test strategies to warrant their implementation. All the Chi-square-based methods previously mentioned are easily implemented using the already existing results of the PSSE process.

However, one major limitation that all the real-life implemented physics-model methods share is that they are quasistatic strategies, meaning they essentially look at a single snapshot in time and analyze only that single snapshot. Especially with the increasing amount of data that comes with the smart grid (SG), it makes sense to develop a strategy that uses not only spatial information (topology and measurements), but temporal information as well (previous measurements). This makes machine learning techniques an obvious candidate, which has led to much research toward applying machine learning to the PSSE state estimation problem.

Considering such, machine learning data-driven linear regression models can be used to estimate measurements in power systems. The motivation for using multiple linear regression is that it is a simple regression model with very few parameters to train. It was shown in (Nagaraj et al., 2020) that a data-driven linear model is capable of achieving satisfying prediction performance on daily load data. Historical measurement values are used as input features to train multiple linear regression models, thus capturing spatial and temporal characteristics. The coefficients estimated using the linear regression model for properly scaled input features provide an easy way to understand the contribution of each input feature in estimating the target.

Considering such, let the number of measurement values be D and the number of past days used as input features for regression models be K. Thus, D linear regression models are trained, corresponding to D measurements. For each model, the corresponding measurements from K past days are used as input features for the regression model.

$$y_d = \beta_d^0 + \beta_d^1 x_{d1} + \ldots + \beta_d^K x_{dK}, \tag{9.3}$$

where the dependent variable $y_d \in \mathbb{R}^{Dx1}$ is a vector that contains d-th measurement values for a current day, the independent variable $x_{dk} \in \mathbb{R}^{Dx1}$ is a vector that contains d-th measurements values from the kth past day, and $\beta_d \in \mathbb{R}^{1x(K+1)}$ is a vector that contains β_d^0, $\beta_d^1 + \ldots + \beta_d^K$ values.

The regression coefficient matrix $\beta \in \mathbb{R}^{Dx(K+1)}$ contains $(K+1)$ values (one for each of the past K days and one intercept) for the D multiple linear regression models. β is estimated by solving (9.1) using the least squares method (Abur and Expósito, 2004). Fig. 9.1 illustrates an example of recorded measurement values and the measurement values predicted by multiple linear regression models trained using past temporal data.

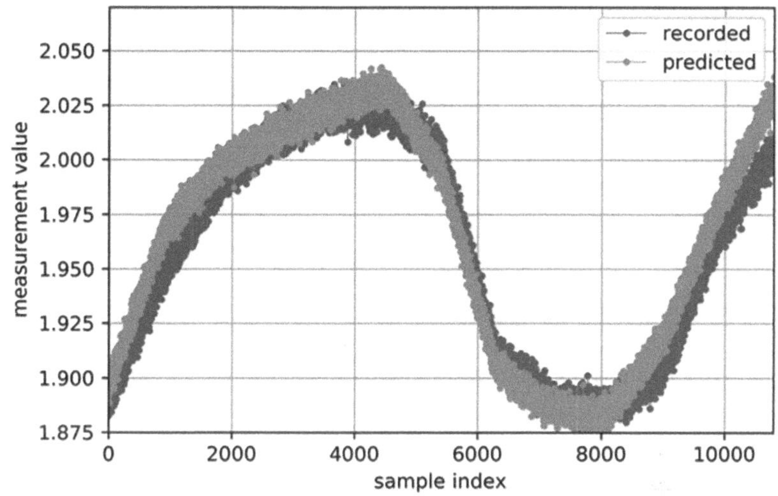

Fig. 9.1 An example for ML predicted measurement vs recorded measurement value.

The values of the weight matrix (W_{ml}) are among the key decisions on the machine learning-based linear regression model. A selection of weight matrices based on regression model coefficients can be used. Considering such, $W_{ml} \in \mathbb{R}^{DxD}$ can be used to weigh the measurements in the state estimation process, and it can be calculated as:

$$W_{ml}^{i,j} = \begin{cases} \dfrac{K+1}{\sum_{k=0}^{K} \left(\beta_i^k - \overline{\beta}_i\right)^2}, & \text{if } i = j \\ 0, & \text{if } i \neq j \end{cases} \tag{9.4}$$

where $i, j = 1, 2, 3, \ldots, D$ represents the measurement number and $\overline{\beta}_i$ represents the average of $K+1$ regression coefficients of measurement i.

After *BDD*, the natural next step for the physics-based model continuous update is to identify the source of error, that is, the bad data identification (*BDI*). There are several presented solutions for BDI. The most recent is the largest normalized error test (LNET), in order to identify gross errors in static and dynamic data, thus extending the largest normalized residual test (Monticelli, 1999). Considering such extension to the largest normalized residual test, Bretas et al. (2019) proved that assuming all the measurements of a set with limited random errors, and adding gross error only to one of them, the measurement to which gross error was added will have the largest increment of error among all the measurements. For such, Bretas et al. (2019) proved that the parameter error spreads out to the measurements associated with the parameter.

However, there are additive difficulties. Depending on the error value, type, and location, the LNET alone might not be able to identify the source of the error. Bretas et al. (2019) has proved that a parameter error will spread to the measurements this parameter is dependent upon, which means that several measurements might have similar normalized errors that might be mistaken by multiple measurement errors.

The question we would like to answer then is how to identify gross errors on dynamic and static data, which can occur simultaneously, while leveraging DRL analyses? Depending on the bad data location, type, and value, it will spread out to the measurement vector space differently. We theorize that the error characteristic will generate different patterns in the measurement vector space. This assumption is plausible because we have proved that the parameter error spreads to all measurements it is related to Bretas et al. (2019).

Thus, if DRL is capable of providing measurement vector space pattern recognition, one might just have the necessary additive information for BDI. Once one has such DRL information, the question we would then like to answer is how to enhance the physics-based model analytic identification test. Fuzzification of the outputs and information fusion might just provide the enhanced identification test. Adaptive threshold values for information fusion can be investigated for generalization.

After BDI, the following step for the continuous update of the physics-based model is bad data correction (BDC). For BDC, considering an error on parameters, one can investigate initially the use of the augmented state vector approach. On this technique, one includes the identified parameter with error, on the state variable vector, and runs the state estimation. While theoretically providing good results, we foresee some difficulties. This approach will be limited by the necessary local redundancy level. Initial results (Bretas et al., 2015; Bretas and Martins, 2015) show that if local redundancy is not sufficient, one will have an unobservable system, thus we will not be able to estimate and thus correct the parameter in error.

For such conditions, one can make use of the composed normalized error (CNE) for correction. The CNE can be used to correct the error. Initial tests on such a correction solution have shown interesting results (Bretas et al., 2017c). However, Bretas et al. (2017c) has considered that all parameters are proportionally equal in error. One solution investigated for such conditions is a phase component approach. In this approach, parameter errors are calculated per phase. In such cases, the measurement model will be developed on phase component representation. Initial results for measurement error correction using the model are encouraging (Bretas et al., 2017b; Zou et al., 2019).

9.2 Ensemble CorrDet with adaptive statistics

Considering the state estimation framework, a most important post processing process is bad data analytics. Several machine learning methods have been developed to detect bad data, model for most FDI cyberattacks in the literature. Some methods rely on neural network-based approaches to identify FDI samples, such as using a simple neural network architecture (Potluri et al., 2017) or deep belief networks (He et al., 2017; Wei et al., 2018).

Neural network-based methods are capable of learning a nonlinear classifier to detect abnormal samples; however, the high cost of computation complexity, especially when the network is deep, might hinder real-life applications. Ozay et al. (2016) investigated some of the commonly used machine learning methods, such as *K*-nearest neighbor (KNN), support vector machine, and sparse logistic regression, to detect FDI samples. However, only temporal information of data is considered in these models while spatial relationship is ignored.

The authors in Xue et al. (2019) proposed the ELM-based one-class-one-network framework to detect FDI attacks and test the solution considering the *IEEE*-14 bus system. Xue et al. (2019) considers a distributed approach by having a separate deep learning model for each bus and a global layer for final decision making, which enables the framework to detect multiple simultaneous FDI attacks. A deep learning model for each bus makes this framework data and computation intense. Although this framework uses spatial relationships in the grid, it fails to capture temporal variations in the normal and abnormal data as the model does not update over time.

A modified KNN-based method for FDI detection was presented in Krundyshev and Kalinin (2019), considering a smart infrastructure setting. Krundyshev and Kalinin (2019) focused on tests on data collected from sensors placed inside a single building, thus avoiding computation issues. A comparative analysis of the modified KNN with other machine learning approaches was implemented using accuracy metrics. Krundyshev and Kalinin (2019) has high computational cost when the number of sensors increases due to the usage of the KNN-based algorithm.

Ashrafuzzaman et al. (2018) presents a deep learning-based model for FDI attack detection using a multilayer perceptron neural network (MLPNN), and compares the attack detection capabilities of their scheme with other machine learning approaches using the *IEEE*-14 bus system. The proposed scheme fails to capture both spatial correlations of the measurements and temporal variations in the data, even when the load profile used is not realistic. Choosing values for a variety of hyperparameters in the MLPNN model also becomes an issue, as large amounts of training data are needed to select optimal values and model tuning.

The CorrDet anomaly detection algorithm, presented in Ho and Gader (2000) and Ho and Gader (2002), is initialized with a sample mean and covariance matrix for the first k number of incoming samples that are not labeled as anomalies, thus identified as normal. These statistics are used to generate statistics for the underlying distribution of normal samples, and to make a linear prediction of the next sample. In Ho and Gader (2000) and Ho and Gader (2002)), the first k number of incoming samples is used to estimate the sample mean and covariance matrix of the normal samples.

Using these initialized mean and covariance values for normal samples, the squared Mahalanobis distance (Chang and Chiang, 2002) of a new incoming sample can then be calculated. The computed distance measure is compared to a decision threshold. If the distance measure is smaller than the threshold, then a new sample is identified as a normal sample; otherwise, the new sample is flagged as an anomaly.

For a time-varying dataset, the CorrDet (CD) detector can provide a solution to adapt statistics with changing trends. Each time a new sample is detected as normal, mean and covariance matrices can be updated using the Woodbury matrix identity (Alvey et al., 2016; Nagaraj et al., 2020). A CD algorithm with fixed statistics was proposed for FDI detection in a dataset where the load profile was kept constant, but with random white noise added to the system, in Trevizan et al. (2019a). The authors used fixed statistics in this work. The hypothesis was that the load profile of the power system does not change for different buses over time, which is not always the real-life application case.

The CD algorithm with fixed statistics will perform poorly for a dataset that has a drifting load profile, or when different buses in power systems have different local statistics (Nagaraj et al., 2020). A full mean and covariance matrix, estimated during the training phase of the CD algorithm with fixed statistics, fails to capture these dynamics. The Extended CorrDet (ECD) algorithm extends the strategies from the CD algorithm and applies it to a dataset that has a drifting load profile in Ruben et al. (2020b). The drift can be modeled by the Ornstein-Uhlenbeck process, which is a stochastic process similar to a random walk but has a tendency to move back toward the original load. This presents a greater challenge to data-driven solutions because the statistics of the different measurements in the power system will no longer be similar.

The ensemble CorrDet (ECD) algorithm provides a distributed anomaly detection scheme while combining measurements using the spatial information of the power system to form a local anomaly detector that learns normal statistics for each bus based on the concept of the CD algorithm (Nagaraj et al., 2020). An optimal threshold is selected for each local anomaly detector and the results are combined with a physics-based solution to make final decisions on FDI detection. Nagaraj et al. (2020) uses fixed statistics for each local anomaly detector and does not update them after initial training, as the state of different buses in power systems does not change over time in a drifting load dataset.

The ECD algorithm will perform poorly for a dataset that has varying system states over time, such as a common daily load profile, which is more realistic compared to constant load profile or a drifting load profile. Fig. 9.2 illustrates how different measurements such as real power injection, reactive power injection, and voltage magnitude change over time for the drifting load profile and the daily load profile.

One can observe that especially different power measurements have large varying normal statistics over time compared to the drifting load profile, even though the variation voltage magnitude is fairly similar for different load profiles. The daily load profile usually has a measurement set that includes real and reactive power flows, real and reactive power injections, and all voltage magnitudes in the grid.

In previous works (Trevizan et al., 2019a; Ruben et al., 2020b), the anomaly detection threshold is a fixed value estimated from the initialization stage and not updated with new incoming data. Specifically, the standard deviation and the mean of squared Mahalanobis distance values of initial k normal samples are calculated first. The threshold that results in the highest $F1$ score for k training normal samples is selected as optimal for the testing phase for the entire system in Trevizan et al. (2019a) and each local anomaly detector in Ruben et al. (2020b). An approach that integrates machine learning techniques to enable an anomaly detection algorithm with an adaptive response to the real-time environment can clearly improve the solution.

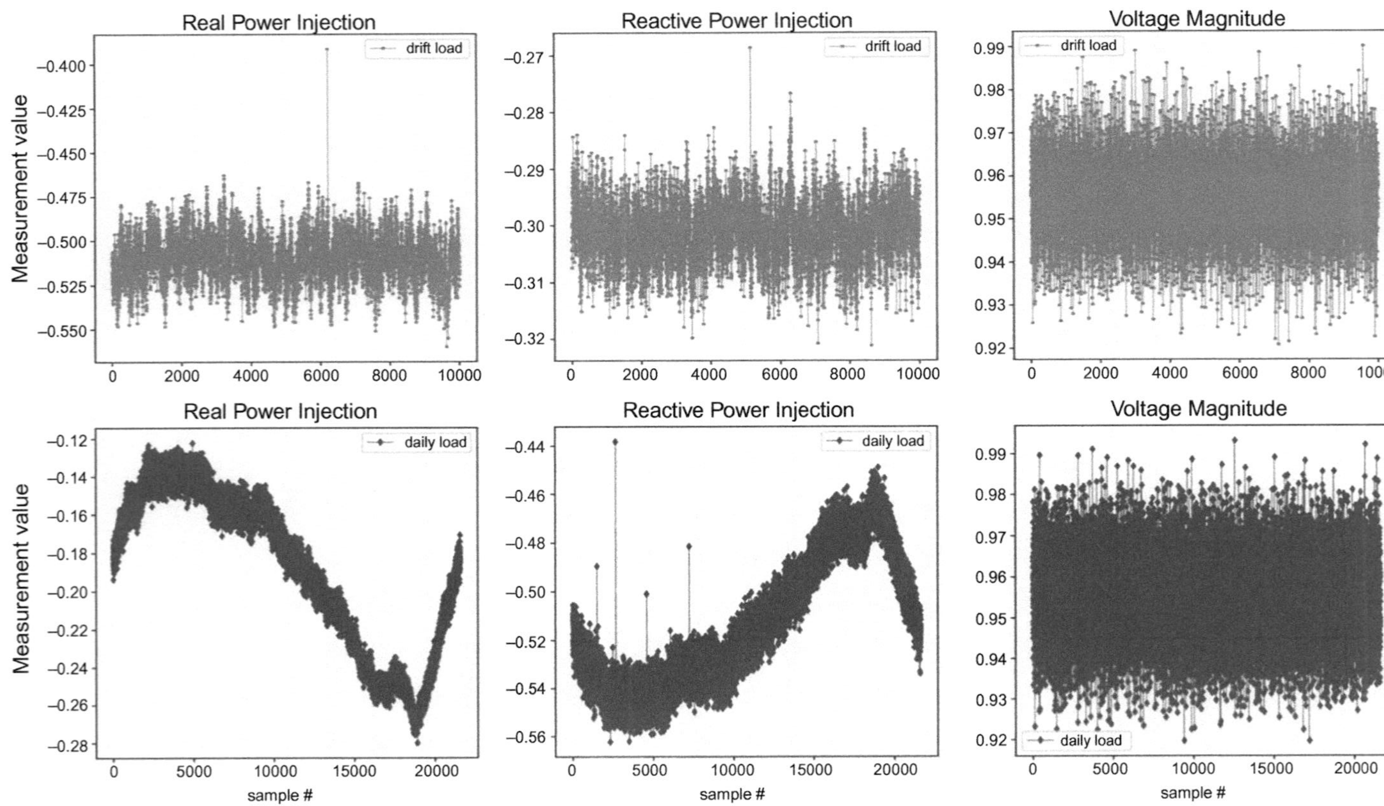

Fig. 9.2 Measurement values for drift load profile (first row plots) and daily load profile datasets (second row plots).

Once the normal state is known (estimated), the system will then be able to detect anomalies introduced into the system as well as alert a network and communications layer (network layer). The network layer can then identify the source of the anomaly, isolate it from the remainder of the system, and take appropriate action to prevent contamination within the connected power distribution systems and within the data assimilation itself.

An ECD with an adaptive statistics (AS) algorithm extends the work of the CorrDet (Trevizan et al. 2019a) algorithm and the ECD algorithm (Ruben et al., 2020b). ECD-AS can be interpreted as a set of CorrDet detectors where adaptive statistics are collected for each local CorrDet environment. The CorrDet anomaly detector (CD) learns a set of statistics, (μ, Σ, and τ) for all buses Φ_R in the power grid topology, and the ECD detector learns a series of statistics (μ_m, Σ_m, and τ_m), one for each bus ϕ_m.

The ECD-AS detectors learn a series of statistics (μ_m, Σ_m, and τ_m), one for each bus, and then processes them with new incoming data samples to adapt them. An adaptive mean and covariance estimation of normal samples is developed as well as an adaptive threshold strategy.

Learning the full covariance over all measurements of all buses in the SG is costly and unnecessary (nearly sparse covariance), especially when training data are limited. In the SG topology, spatially neighboring buses are more highly correlated and can be more easily affected by bad data while buses that are further apart have lower correlation. Data from local buses with fewer dimensional measurement sets offer a more accurate statistic estimation and a computationally cheaper, more sensitive anomaly detection.

Unlike the previous algorithms, ECD-AS prevents the numerical issue of estimating a high-dimensional mean and covariance for the distribution of the normal samples (in CorrDet detector) in the space of all measurements when the number of measurements is high and the number of training samples is low, by learning lower-dimensional statistics in the space of only the measurements associated with each bus.

Let the total number of measurements be d. There are m_j, $m_j < d$, measurements on each bus m, where each bus is considered as a local spatial region, corresponding to one Local CorrDet detector with adaptive statistics, ϕ_m. For Φ_R, the learning process consists of estimating μ and Σ^{-1} from normal training samples z ($z \in \mathbb{R}^{1 \times d}$). A similar strategy can be made to learn the series of statistics for the ECD-AS detector. The learning of Φ_E involves the estimation of a set of Local CorrDet detectors with adaptive statistics, ϕ_m. For each ϕ_m, similarly, the learning process consists of estimating its μ_m and $\sum_m^{-1}$ from the normal training samples with selected measurements z_m (z_m is a $1 \times m_j$ vector).

First, the μ_m and $\sum_m^{-1}$ are initialized with the sample mean and with the covariance of selected measurements of the first k samples that are labelled as normal. Then, the detector starts to accept new samples and classify each new sample as follows. For each new incoming sample z, a set of squared Mahalanobis distances, δ_m^{ECD-AS}, is computed using (9.3):

$$\delta_m^{ECD-AS}(z_m) = (z_m - \mu_m)^T \sum_m^{-1} (z_m - \mu_m), \tag{9.5}$$

where μ_m is the mean and $\sum_m^{-1}$ is the inverse covariance matrix of normal samples on the mth local CorrDet detector. Then, the Mahalanobis distances are compared with the corresponding set of thresholds, T, where $T = \{\tau_m\}$ m = 1: M. If at least one squared Mahalanobis distance in Φ_E is greater than its corresponding threshold, this incoming sample is classified as an anomaly. Otherwise, it is classified as a normal sample.

Furthermore, for each new sample that is classified as normal, the anomaly detector must be able to adapt with changing trends to be able to capture data that are dynamic and change gradually over time. Therefore, the mean, μ_m, and the inverse covariance matrix, $\sum_m^{-1}$, are updated using the Woodbury matrix identity (Alvey et al., 2016; Nagaraj et al., 2020), [43] using (9.4) and (9.5), respectively. Note that this update is done only if the incoming data are considered normal data. The mean will be updated such that

$$\mu_{new,m} = (1-\alpha)\mu_m + \alpha(z_m - \mu_m)' \tag{9.6}$$

where $\alpha \in \mathbb{R}^{1x1}$ is the weight value for the incoming data sample, and the inverse covariance matrix otherwise is updated by:

$$\sum_{new,m}^{-1} = \frac{1}{1-\alpha}\left[\sum_m^{-1} - \frac{(z_m - \mu_m)(z_m - \mu_m)'}{\left(\frac{1-\alpha}{\alpha}\right) + (z_m - \mu_m)'(z_m - \mu_m)}\right] \tag{9.7}$$

where μ_m is the old mean of the mth local CorrDet detector with adaptive statistics, $\sum_m^{-1}$ is the old inverse covariance matrix of the mth local CorrDet detector with adaptive statistics, and α is a hyperparameter value between zero and one that determines how much importance is given to the new data sample versus the old mean.

The values of k and α can be obtained through experimentation. The threshold can be assumed to be fixed for a dataset that has constant mean and small variation in the time domain. A fixed threshold $T=\{\tau_m\}_{m=1:M}$ can be estimated using (9.6):

$$\tau_m = \mu_{thr,m} + \eta\sigma_{thr,m} \tag{9.8}$$

where $\mu_{thr,\ m}$ and $\sigma_{thr,\ m}$ are the mean and standard deviation of the squared Mahalanobis distance values, respectively. These values are estimated only for normal classified training samples, with selected measurements associated with the *m*th local CorrDet detector with adaptive statistics.

The fixed threshold assumption holds on a dataset that has constant mean and small variation in the time domain, such as constant load profile and drifting load profile. However, for the daily load profile, the statistics of normal samples have a larger dynamically changing mean and covariance with time, as shown in Fig. 9.1. This ensures that a fixed threshold assumption does not hold. So, in addition to adaptive mean and covariance estimation of normal samples, an adaptive threshold strategy should be used.

Adaptive threshold estimation initializes the threshold values τ_m for each local CorrDet detector with adaptive statistics (bus-level) following (9.8), and updates τ_m in an online sliding window fashion. For every new incoming sample z, the threshold values τ_m are inferred from the most recent β normal samples before it. In other words, the standard deviation ($\sigma_{thr,m}$) and mean ($\mu_{thr,m}$) of the squared Mahalanobis distance values of β normal samples past the new sample z are calculated for each local CorrDet detector with adaptive statistics ϕ_m. Here, β is the sliding window size.

The threshold value τ_m for each local CorrDet detector with adaptive statistics is updated using (9.9) with updated $\mu_{thr,m,-\beta}$ and $\sigma_{thr,m,-\beta}$ where $-\beta$ means the use of the past β number of samples for updating the threshold:

$$\tau_m = \mu_{thr,m,-\beta} + \eta\sigma_{thr,m,-\beta}. \tag{9.9}$$

References

Abur, A., Expósito, A.G., 2004. Power system state estimation: theory and implementation. In: Power Engineering (Willis). CRC Press, USA. Available from: https://books.google.com/books?id=NQhbtFC6_40C.

Alvey, B., Zare, A., Cook, M., et al., 2016. Adaptive coherence estimator (ACE) for explosive hazard detection using wideband electromagnetic induction (WEMI). ProcSPIE 9823, 9823. Available from: https://doi.org/10.1117/12.2223347.

Ashrafuzzaman, M., Chakhchoukh, Y., Jillepalli, A.A., et al., 2018. Detecting stealthy false data injection attacks in power grids using deep learning. In: 2018 14th Int. Wireless Communications Mobile Computing Conf. (IWCMC), Limassol, Cyprus, pp. 219–225.

Bobba, R.B., Rogers, K.M., Wang, Q., et al., 2010. Detecting false data injection attacks on dc state estimation. In: Preprints of the First Workshop on Secure Control Systems, CPSWEEK, Stockholm, Sweden, 2010.
Bretas, N.G., Bretas, A.S., 2018. The extension of the Gauss approach for the solution of an overdetermined set of algebraic non linear equations. IEEE Trans. Circ. Syst II Exp. Briefs 65 (9), 1269–1273.
Bretas, N.G., Martins, A.C.P., 2015. Transmission line parameter error detection, identification and correction with geometrical view: Topological errors. In: 2015 IEEE Eindhoven PowerTech, June 2015, pp. 1–4.
Bretas, N.G., Bretas, A.S., Piereti, S.A., 2011. Innovation concept for measurement gross error detection and identification in power system state estimation. IET Gener. Transm. Distrib. 5 (6), 603–608.
Bretas, N.G., Piereti, S.A., Bretas, A.S., et al., 2013a. A geometrical view for multiple gross errors detection, identification, and correction in power system state estimation. IEEE Trans. Power Syst. 28 (3), 2128–2135.
Bretas, N.G., Bretas, A.S., Martins, A.C.P., 2013b. Convergence property of the measurement gross error correction in power system state estimation, using geometrical background. IEEE Trans. Power Syst. 28 (4), 3729–3736.
Bretas, N.G., Carvalho, B.E.B., Albertini, M.R.M.C., 2015. The innovation concept applied to the processing of measurements and parameters errors in power systems state estimation. In: 2015 IEEE Eindhoven PowerTech, June 2015, pp. 1–6.
Bretas, A.S., Bretas, N.G., Carvalho, B., et al., 2017a. Smart grids cyber-physical security as a malicious data attack: an innovation approach. Electr. Power Syst. Res. 149, 210–219. Available from: http://www.sciencedirect.com/science/article/pii/S0378779617301657.
A. Bretas, N. Bretas, S. Braunstein, A. Rossoni, and R. Trevizan, "Multiple gross errors detection, identification and correction in three-phase distribution systems wls state estimation: a per-phase measurement error approach," Electr. Power Syst. Res., vol. 151, pp. 174 – 185, 2017b. [Online]. Available: http://www.sciencedirect.com/science/article/pii/S0378779617302109.
Bretas, A.S., Bretas, N.G., Carvalho, B., Baeyens, E., Khargonekar, P.P., 2017c. Smart grids cyber-physical security as a malicious data attack: an innovation approach. In: Electric Power Systems Research, vol. 149, pp. 210–219. [Online]. Available: http://www.sciencedirect.com/science/article/pii/S0378779617301657.
Bretas, A.S., Bretas, N.G., Carvalho, B.E.B., 2019. Further contributions to smart grids cyber-physical security as a malicious data attack: proof and properties of the parameter error spreading out to the measurements and a relaxed correction model. Int. J. Electr. Power Energy Syst. 104, 43–51. Available from: http://www.sciencedirect.com/science/article/pii/S0142061518303946.
Chang, C.I., Chiang, S.S., 2002. Anomaly detection and classification for hyperspectral imagery. IEEE Trans. Geosci. Remote Sens. 40 (6), 1314–1325.
Choquet, G., 1954. Theory of capacities. Ann. Inst. Four. 5, 131–295. [Online]. Available: https://aif.centre-mersenne.org/item/AIF_1954__5__131_0.
G. Ferreira, D. Gazzana, R. Silva, A. Bretas, A. Bettiol, A. Carniato, L. Passos, R. Homma, and F. Molina, "Analytical goal programming model for optimal restoration of distribution systems," IEEE Power and Energy Society General Meeting, vol. 2015 2015. [On-line]. Available: https://www.scopus.com/inward/record.uri?eid=2-s2.0-84956852435&doi=10.1109%2fPESGM.2015.7286036&partnerID=40&md5=0196999533942c83243fefa54f1687e8.
Handschin, E., Schweppe, F.C., Kohlas, J., et al., 1975. Bad data analysis for power system state estimation. IEEE Trans. Power Appar. Syst. 94 (2), 329–337.
He, Y., Mendis, G.J., Wei, J., 2017. Real-time detection of false data injection attacks in smart grid: a deep learning-based intelligent mechanism. IEEE Trans. Smart Grid 8 (5), 2505–2516.

Ho, K.C., Gader, P.D., 2000. Correlation-based land mine detection using GPR. Detection and remediation technologies for mines and minelike targets V. Int. Soc. Opt. Photon. 4038, 1088–1096. Orlando, FL, USA.

Ho, K., Gader, P.D., 2002. A linear prediction land mine detection algorithm for hand held ground penetrating radar. IEEE Trans. Geosci. Remote Sens. 40 (6), 1374–1384.

Kosut, O., Jia, L., Thomas, R.J., et al., 2010. Malicious data attacks on smart grid state estimation: attack strategies and counter measures. In: 2010 First IEEE Int. Conf. on Smart Grid Communications, Gaithersburg, MD, USA, pp. 220–225.

Krundyshev, V., Kalinin, M., 2019. Prevention of false data injections in smart infrastructures. In: 2019 IEEE Int. Black Sea Conf. on Communications and Networking (BlackSeaCom), Sochi, Russia, pp. 1–5.

Monticelli, A., 1999. State estimation in electric power systems: a generalized approach. 507 Springer Science & Business Media, USA.

Nagaraj, K., Zou, S., Ruben, C., Dhulipala, S., Starke, A., Bretas, A.S., McNair, J., Zare, A., 2020. Ensemble CorrDet with adaptive statistics for bad data detection. IET Smart Grid 3, 572–580. https://doi.org/10.1049/iet-stg.2020.0029.

Ozay, M., Esnaola, I., Vural, Y.F.T., et al., 2016. Machine learning methods for attack detection in the smart grid. IEEE Trans. Neural Netw. Learning Syst. 27 (8), 1773–1786.

Potluri, S., Diedrich, C., Sangala, G.K.R., 2017. Identifying false data injection attacks in industrial control systems using artificial neural networks. In: 2017 22nd IEEE Int. Conf. on Emerging Technologies and Factory Automation (ETFA), Limassol, Cyprus, pp. 1–8.

Rossoni, A., Braunstein, S., Daniel Trevizan, R., Bretas, A., Bretas, N., 2018. In: Contribution to Distribution Systems Technical and Nontechnical Losses Estimation Using wls State Estimator. 2017 IEEE Power and Energy Society General Meeting, vol. 2018-January, pp. 1–5. [Online] Available https://www.scopus.com/inward/record.uri?eid=2-s2.0-85046335642&doi= 10.1109%2fPESGM.2017.8273804&partnerID=40&md5=e55ce321c5a9b6e8c3dc82fd1a45d07b.

Ruben, C., Dhulipala, S., Bretas, A.S., Guan, Y., Bretas, N.G., 2020a. Multi-objective MILP model for PMU allocation considering enhanced gross error detection: a weighted goal programming framework. Electr. Power Syst. Res. 182, 1–10. https://doi.org/10.1016/j.epsr.2020.106235.

Ruben, C., Dhulipala, S.C., Nagaraj, K., et al., 2020b. Hybrid data-driven physics model-based framework for enhanced cyber-physical smart grid security. IET Smart Grid, Available from: https://digital-library.theiet.org/content/journals/ https://doi.org/10.1049/iet-stg.201.

Trevizan, R., Ayar, M., Bretas, A., Obuz, S., 2018. Cooperative control of energy storage for transient stability enhancement. In: IEEE Power and Energy Society General Meeting. vol. 2018. [Online]. Available https://www.scopus.com/inward/record.uri?eid=2-s2.0-85060808644&doi=10.1109%2fPESGM.2018.8586028&partnerID=40&md5=5bec22d5716f6e74890d8b7dbe4d52d8.

Trevizan, R.D., Ruben, C., Nagaraj, K., et al., 2019a. Data-driven physics-based solution for false data injection diagnosis in smart grids. In: IEEE PES GM, Atlanta, GA, USA, 2019.

Trevizan, R., Ruben, C., Nagaraj, K., Ibukun, L., Starke, A., Bretas, A., McNair, J., Zare, A., 2019b. Data-driven physics-based solution for false data injection diagnosis in smart grids. In: IEEE Power and Energy Society General Meeting.

Wei, L., Gao, D., Luo, C., 2018. False data injection attacks detection with deep belief networks in smart grid. In: 2018 Chinese Automation Congress (CAC). Xi'an, China, pp. 2621–2625.

Xue, D., Jing, X., Liu, H., 2019. Detection of false data injection attacks in smart grid utilizing ELM-based OCON framework. IEEE Access 7, 31762–31773.

Zou, T., Bretas, A., Aljohani, N., Bretas, N., 2019. Malicious data injection attacks: a relaxed physics model based strategy for real-time monitoring. In: Proceedings of the 51st North American Power Symposium.

Index

Note: Page numbers followed by *f* indicate figures, *t* indicate tables, and *b* indicate boxes.

FSC
www.fsc.org
MIX
Papier | Fördert gute Waldnutzung
FSC® C083411

Kolibri 360 GmbH
Ferdinand-Jühlke-Straße 7
99095 Erfurt, Deutschland
produktsicherheit@kolibri360.de